DETAIL Practice

Acoustics and Sound Insulation

Principles
Planning
Examples

Eckard Mommertz
Müller-BBM

Birkhäuser
Edition Detail

Author:
Eckard Mommertz, Dr.-Ing.
Müller-BBM, Planegg

Assistants:
Gunter Engel, Dipl.-Phys., Dipl.-Tonmeister
Martina Freytag, Dipl.-Ing.
Gerhard Hilz, Dipl.-Ing.
Andreas Meier, Dr.-Ing.
Michael Prüfer, Dipl.-Ing.
Elmar Schröder, Dipl.-Phys.
Alexander Schröter, Dipl.-Ing.

All the contributors to this book are employees of Müller-BBM, a firm of consulting engineers with more than 250 employees which specialises in the areas of buildings, environment and technology.
Müller-BBM has provided acoustics and other services for numerous projects over many years, and the wealth of experience gained was incorporated in this book. Among those projects are all those depicted here in the form of photographs and/or examples.

Editorial services:
Melanie Schmid, Dipl.-Ing.

Editorial assistants:
Nicola Kollmann, Dipl.-Ing.; Marion Linssen; Florian Metzeler

Drawings:
Caroline Hörger, Dipl.-Ing.; Daniel Hajduk, Dipl.-Ing.

Translators (German/English):
Gerd H. Söffker, Philip Thrift, Hannover

© 2008 Institut für internationale
Architektur-Dokumentation GmbH & Co. KG, Munich
An Edition DETAIL book

ISBN: 978-3-7643-9953-5
Printed on acid-free paper made from cellulose bleached without the use of chlorine.

Typesetting & production:
Simone Soesters

Printed by:
Aumüller Druck, Regensburg
1st edition, 2009

This book is also available in a German language edition (ISBN 978-3-920034-23-2).

A CIP catalogue record for this book is available from the Library of Congress, Washington D.C., USA.

Bibliographic information published by
Die Deutsche Bibliothek
Die Deutsche Bibliothek lists this publication in the Deutsche Nationalbibliographie; detailed bibliographic data is available on the internet at http://dnb.ddb.de.

Institut für internationale
Architektur-Dokumentation GmbH & Co. KG
Sonnenstraße 17, D-80331 München
Telefon: +49/89/38 16 20-0
Telefax: +49/89/39 86 70
www.detail.de

Distribution Partner:
Birkhäuser – Publishers for Architecture
PO Box 133, 4010 Basel, Switzerland
Tel.: +41 61 2050707
Fax: +41 61 2050792
e-mail: sales@birkhauser.ch
www.birkhauser.ch

Books are to be returned on or before
the last date below.

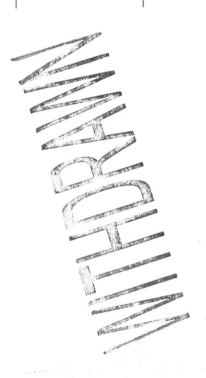

DETAIL Practice
Acoustics and Sound Insulation

Contents

6   *Introduction*

8   *Designation and auditive perception of sound*

12   *Room acoustics*

24   *Building acoustics*

38   *Noise control in urban planning*

48   *Sound insulation as a quality feature in housing*
56   Case study: Inner-city development, Munich

58   *Office buildings*
66   Case study: Swiss Re offices, Munich

68   *Schools and preschool facilities*
75   Case study: Primary school in Erding

78   *Lecture theatres, congress halls, plenary chambers*
80   Case study: Zollverein School of Management and Design in Essen

82   *Small rooms for music*
85   Case study: Music school in Grünwald
88   Case study: Conversion of an officers' casino into a music school in Landshut

90   *Rooms for sound*
103   Case study: Opera house in Hangzhou
104   Case study: Philharmonic in Essen

106   *Churches*

Appendix
108   Authorities, institutes and trade associations, bibliography,
      standards and directives, manufacturers
111   Index
112   Picture credits

Introduction

"Acoustics" is derived from the Greek word ακουειν (akouein), which means "to hear", and is the branch of science that deals with sound, including its generation, transmission, analysis and perception. More than 2500 years ago, Pythagoras investigated musical relationships, and in architecture, Vitruvius (c.70–10 BC) describes the acoustic design of amphitheatres. In the 19th century, acoustics advanced to become a scientific discipline, which since the early 20th century has also included the acoustics of rooms and buildings.

These days, the expression "good acoustics" is mostly associated with famous concert halls, possibly also ancient amphitheatres. Actually, though, every building, every room has an acoustic dimension. Hearing and understanding are fundamental prerequisites for communication, and the acoustic feedback of an interior for speech or music is essential; infiltrating noise is disturbing – and can even be unhealthy.

In the majority of cases, the auditive effect of an interior space is not perceived consciously. But this changes when the aural impression does not meet our expectations: difficulty in understanding presentations in a seminar room, distracting noises in an open-plan office, or poor sound insulation between adjacent apartments.

We distinguish between room acoustics and building acoustics. The former deals with the sound in a room, i.e. how the shape, size and surfaces in the room influence the physics of sound propagation and how this might affect our auditory perception. In the end, it is the room acoustics that determine how good an interior space functions for spoken communications, or for different types of music. "Catastrophic room acoustics" in the form of reverberating, loud rooms can generally be avoided by designing properly, provided this topic is taken into account in the architectural design in good time. Good design plus – especially where music is involved – experience and intuition are the keys to achieving good room acoustics.

Building acoustics, on the other hand, is primarily concerned with preventing sound propagation within the building, in order to avoid the spread of disturbing noise. The arrangement of different functions within the building and the appropriate acoustic design of constructions and components are the essential aspects here.

This book is intended for architects, building owners, developers, engineers and all those interested in the theme of acoustics in buildings.

The first part of the book covers the principles, relationships, international and German standards plus planning and prognosis methods for the subjects of room acoustics, building acoustics and noise control in urban situations.

This is followed by chapters covering various types of building, types of usage, which includes residential and office buildings, schools and preschool facilities, lecture theatres, congress halls, plenary chambers, small rooms for music, cultural venues and churches.

The principles of acoustics are reinforced by way of case studies.

These chapters can provide a guide for specific projects and draw attention to the acoustic aspects of a construction project in good time. Advice regarding solutions or potential solutions can be found here, but these are not "patent recipes" – the issues involved are always unique to the specific project.

And finally, this book is intended to heighten the reader's awareness of the fact that appropriate acoustic conditions can enhance the success of a project.

Designation and auditive perception of sound

Sound is understood to be the small, fast pressure and density fluctuations that are caused by vibrating or fast-moving objects. Examples of this are the rustling of leaves in the wind, the rippling of a stream, the oscillating string of a musical instrument, or the vibrations of an engine. Beginning at the source, the sound propagates in all directions at a velocity of about $c = 340$ m/s (or 1200 km/h) in air, the speed of sound. As it propagates it is influenced in various ways, e.g. amplified by numerous reflections in a room, or weakened upon passing through a wall. Once it arrives at the human ear, the sound causes the eardrum to vibrate. These vibrations are transmitted via the hammer, anvil and stirrup, the tiny bones of the middle ear, with a sort of lever action, to the inner ear, where together with other sensory stimuli they can trigger the most diverse feelings and reactions.

Sound and noise
In principle, we distinguish between two categories of sound.
Firstly, sound that serves communications or our well-being, e.g. listening to and playing music, speaking and listening. These sound signals should be transmitted in such a way that a good timbre is achieved and speech is understandable – the principal tasks of room acoustics.

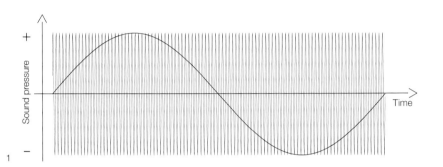

1

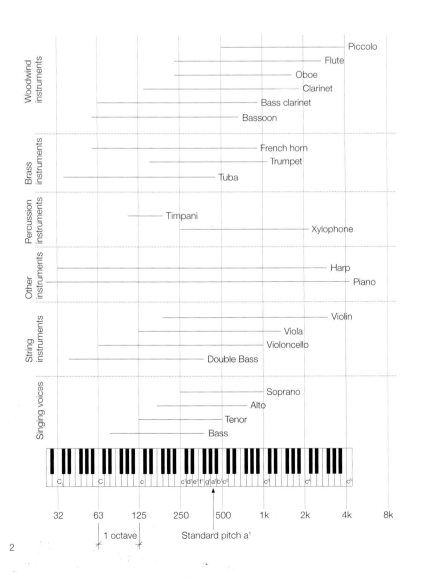

1   Two sinusoidal pitches (pure tones) with frequencies that differ by a factor of 100. High pitches correspond to short wavelengths, low pitches to long wavelengths.
A sound, noise or even a bang are all made up of a multitude of sine-waves together, the composition of which changes over time.
2   Frequency ranges of musical instruments and singing voices

2

T1: Physical description and auditive perception

Frequency	Pitch
Sound pressure level	Loudness
Combination of frequencies	Timbre

The opposite to this is sound that is a nuisance and can even be unhealthy, e.g. traffic or aircraft noise infiltrating into an apartment, the booming base of a neighbour's hi-fi system. Such problems call for constructional measures to reduce the noise level, taking into account regulations and legislation.

Airborne and structure-borne sound
Sound cannot propagate in a vacuum. Sound requires a medium for its transmission and we distinguish between airborne and structure-borne sound. The speed of sound in solid bodies is about 5 – 15 times faster than that in air, depending on the material, and, furthermore, far more complex because different wave forms ensue.

Frequency and wavelength
The sounds audible by the human ear range over a broad spectrum of frequencies from about 20 to 20 000 Hz. This frequency spectrum extends downwards into the infrasound range and upwards into the ultrasound range.
The following relationship links wavelength λ and frequency f:

$$\lambda = c/f$$

Consequently, the audible range of wavelengths ranges from approx. 10 m down to approx. 20 mm, i.e. a ratio of 1:500. Compared with this, the wavelengths of visible light (electromagnetic waves) between infrared and ultraviolet differ by only a factor of two (380 to 750 nm). Sound signals are mostly concordant (sound) or discordant (noise) combina-tions of diverse frequencies. Even when depressing a key on the piano it is not one frequency that is generated but always a fundamental tone plus additional over-tones (harmonics). However, the frequency of the fundamental tone is generally responsible for the sound we perceive subjectively.

Fig. 2 shows the typical frequency ranges of singing voices and some musical instruments.
We combine frequencies into so-called frequency bands for the purpose of measuring, analysing or even specifying the acoustic properties of building com-ponents. The usual classification is to employ octaves with centre frequencies of 125 Hz, 250 Hz, 500 Hz, 2000 Hz and 4000 Hz. An octave band covers a fre-quency ratio of 1:2. Smaller frequency intervals also in use are one-third octave bands, i.e. thirds, (frequency ratio 1:1.28). In music, an octave is divided into 12 semitones, as is reflected in the layout of, for example, a piano keyboard.

Sound pressure and loudness level
The physical variable for designating the strength of a sound is the sound pressure p. This pressure, changing over time with the frequency, ensues due to the particles of air being moved out of their state of rest and is superimposed on the atmos-pheric pressure.

Our auditive perception capacity embraces an enormous dynamic: a factor of 1:1 000 000 relates the threshold of hear-ing (20 µPa) to the threshold of pain (20 Pa). Easier to understand is the logarithmic presentation as the sound pressure level Lp, which in addition tends to correspond to our subjective perception of loudness. The sound pressure level is specified in decibels (dB) and is calculated from the sound pressure p as follows:

$$L_p = 10 \cdot \log p^2/p_0^2$$

The reference sound pressure $p_0 = 20$ µPa corresponds to the threshold of hearing at medium frequencies. As a comparison, the atmospheric pressure – approx. 100 000 Pa (= 1000 mbar) – is five billion times higher.
It is not the sound pressure p, but rather the square of the sound pressure p^2 that we take as our energy variable. When we speak of sound energy or sound intensity, these variables are generally proportional to p^2. If the volume of traffic on a road doubles, for instance, p^2 also doubles. The equation reveals that doubling the energy corresponds to a 3 dB increase in the sound pressure level.

T2: Sound pressure and sound pressure level for typical acoustic environments

	Sound pressure p [Pa]	Sound pressure level L_{pA} [dB]	
■	20.0	120	Propeller aircraft take-off, threshold of pain
▨	2.0	100	Pneumatic drill, discotheque
▨	0.2	80	Shouting, busy road
▨	0.02	60	Normal speech, loud dishwasher
▨	0.002	40	Whispering, mechanical ventilation in offices
▨	0.0002	20	Bedroom in quiet area, recording studio
□	0.00002	0	Threshold of hearing

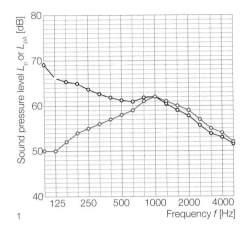

1 Frequency-related sound pressure level (in third octave band) for a busy road
Linear: physically measured level
A-weighted: level adjusted to correspond to human hearing; the difference in levels between the two curves corresponds to the A-weighting. The total level is calculated by determining the squares of the sound pressure from the third octave band levels and adding these together.

o—o Linear, L_p = 75 dB
o—o A-weighted, L_{pA} = 70 dB

Subjectively, though, a 10 dB increase in the sound pressure level is perceived as being "twice as loud", although this corresponds to a 10-fold increase in the sound energy. A subjective four-fold increase in the loudness level calls for a difference in levels of 20 dB; a difference of 3 dB (doubling or halving the sound energy) is perceptible, whereas a difference of 1 dB can only be detected in a direct comparison (p. 9, Tab. T2).

These relationships also help to estimate the reasonableness of noise control measures. For instance, if major construction work is required to reduce a disturbing noise level by 1 dB, this is certainly not justified unless required by legislation.

A-weighting

The human ear is most sensitive to medium frequencies. This sensitivity decreases markedly with very low or very high frequencies. In addition, our hearing ability for high frequencies depends very much on age, but also on "what our ears are accustomed to" (frequent discotheque visits, loud working conditions, etc.). Pitches > 16 kHz are frequently no longer heard by adults.

The A-weighting takes into account this frequency-related sensitivity. This evaluation is achieved by designing the measuring instruments to weight the energy components in the different frequency bands (e.g. third octave band) according to the sensitivity of the human ear. The outcome is the A-weighted sound pressure level L_{pA}, recognised as such by the subscript A (Fig. 1). If sounds are described by way of their sound pressure level, it is usually the A-weighted level that is meant, even if this is not explicitly mentioned. Typical A-weighted sound pressure levels for everyday sounds can be found in Tab. T2 (p. 9).

In addition to the A-weighting, there is also the C-weighting, which is occasionally used for assessing low-frequency noise immissions. The D-weighting is used exclusively for aircraft noise. The different weighting curves take into account the fact that the frequency-related sensitivity of the human ear depends on the level.

Effects of sounds

But how sound is perceived in the end in no way depends solely on the A-weighted sound pressure level, but rather on many other, sometimes also subjective, factors. And this applies not only to the subjective loudness, but to a greater extent to the perceived nuisance level. Psychoacoustics is dedicated to studying such issues, but these are very complex and up until now have hardly affected the design of buildings, or construction standards and regulations. In the design of cars, vacuum cleaners and many more everyday items, however, much more effort is placed on trying to convey a feeling of quality by way of acoustic product design. Examples from the automotive industry are doors that close with a reassuring thud, which are preferred to those that slam, and the light switches that due to weight-savings sound too "cheap".

Despite the complex relationships, different disturbing noises can be described – simplified – in quality terms as follows: levels as low as about 35 dB(A) can disrupt sleep, communication can be impaired from about 45 dB(A) upwards, and our ability to concentrate starts to diminish above a level of about 55 dB(A). If we are exposed to noise levels exceeding 85 dB(A) for long periods, damage to our hearing will be the result. The threshold of pain for noise is around 120 dB(A). In order to enable untroubled spoken communications, the signal level, i.e. the sound pressure level of the speech, should lie at least 10 dB, preferably 15 dB,

above the ambient noise level of, for example, a ventilation system or road traffic. Everyone is familiar with the situation where a speaker is difficult to understand because he or she speaks too quietly. But we can look at this a different way: it is not the speaker who is too quiet, but the ambient noise level that is too loud.

In accordance with these fundamental relationships, we can place demands on the maximum permissible noise levels in rooms – depending on their usage – in order to avoid health risks or impair the function of the room. The ambient noise level may be caused by the infiltration of road traffic noise, building services, toilets, etc. These aspects are explored elsewhere in this book.

Duration of sounds

The description of sound signals above says nothing about their duration.

It is frequently the case that an average sound pressure level, e.g. during a working day, is considered for a certain period of observation, which is expressed by the equivalent continuous sound pressure level L_{Aeq}. If the alleged disturbing effect is based on noise peaks, the maximum level $L_{AF,max}$ is often used. This aspect is investigated in more detail on p. 42, in the chapter entitled "Noise control in urban planning".

Depending on the application, there are many further level definitions which occasionally can be very confusing. However, such distinctions are indispensable when defining guidance or limit values relevant to immissions legislation.

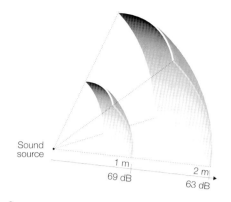

2

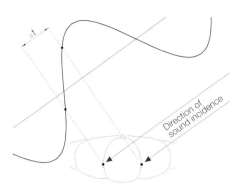

3

T3: Audibility of differences in level

Difference in levels	Audibility
1 dB	Just audible
3 dB	Audible
5 dB	Clearly audible
10 dB	Subjective doubling of loudness
20 dB	Subjective four-fold increase in loudness

T4: Sound power levels of musical instruments and singing voices

Sound source	Sound power level L_{WA} [dB] pp–ff
String instruments	55–95
Woodwind instruments	70–100
Brass instruments	70–115
Grand piano	70–115
Percussion instruments	90–120
Organ, whole orchestra	up to 135
Singing voices	80–115

Designation of sound sources
Technical sound sources can be designated by way of the sound power P (in watts) they emit or by their sound power level L_W. For instance, these days even gardening implements and household appliances such as lawnmowers, refrigerators, etc. are increasingly being categorised via their A-weighted sound power level L_{WA} within the scope of the CE marking system. And the sound power level from pianissimo (*pp*) to fortissimo (*ff*) for musical instruments can also be expressed in terms of decibels (Tab. T4).
The sound power level serves exclusively to define sources and does not express the sound pressure level at the place where the sound is heard.

Sound propagation in open air
Sound radiates from a source in all directions. So as the distance r from the source increases, the sound intensity is distributed over the ever-enlarging sur-face area of an imaginary sphere with radius r, i.e. $4\pi r^2$ (Fig. 2). For non-directional sources, the relationship between sound power level and sound pressure level can be expressed as follows:

$$L_p = L_W - 10 \cdot \log 4\pi r^2$$

where r is expressed in m.

Accordingly, at a distance of 1 m from the sound source, in terms of numbers the sound pressure level is 11 dB below the sound power level. These relationships apply to larger distances from the source, in other words for distances greater than the largest dimension of the source.

Directional characteristic
Generally, sound energy is not radiated evenly in all directions. For example, when speaking, the A-weighted sound pressure level in the direction of speaking is about 5 dB higher than that measured at the same distance from the source but in the opposite direction. Loudspeakers, too, exhibit a distinct directional charac-teristic. We can exploit this fact in order to direct the sound into an audience and minimise exciting the room itself.
As a rule, the directional effect increases with higher frequencies. At frequencies with a wavelength greater than the dimen-sions of the sound source, the radiation of the sound is always non-directional.

Aural determination of direction
Besides the perception of different pitches, timbres and the change in the sound over time, the human ear can also localise the sound – even with our eyes closed. One reason for this is our ears, which filter the sound differently depending on the direc-tion of incidence. In addition, if sound sig-nals come from the side, this leads to an interaural time difference $\triangle t$ between our ears. The associated phase differences are smaller for low frequencies (long wavelengths), which is why low pitches are harder to localise (Fig. 3). This fact enables us to separate low-frequency subwoofers from mid-range and treble speakers in hi-fi and even large sound reinforcement systems.
A sound recording suitable for the human ear can be achieved with the help of a so-called dummy head. This is a more or less faithful reproduction of a human head in which microphones are positioned at the ends of the artificial ears to represent the eardrums. Dummy heads are primarily used for metrology purposes; they are less suitable for high-quality live record-ings of concerts, for instance, because the sounds experienced in a living room differ too much from those experienced in the concert hall.

2 The segment from a sphere makes the situation clear:
Doubling the distance means that the sound energy is distributed over an area four times as large. The sound pressure level decreases by 6 dB accord-ingly. The figures given here are valid for a sound power level at the source of L_W = 80 dB.

3 Aural determination of direction: If the sound comes from the sides, it reaches one ear earlier than the other. The associated frequency-related phase differences are evaluated by our ears. But owing to the long wavelengths at low frequencies, it is difficult to localise low pitches (250 Hz).

11

Room acoustics

The expression "good acoustics" is mostly associated with concert halls, opera house and theatres. Without doubt, good acoustics are vital for such venues. But in the end all interior spaces in which communication takes place by way of speech or music benefit from the right acoustic environment. If speech cannot be understood properly in a lecture theatre or classroom, this is often a great problem. And in rooms for musical performances or rehearsals, acoustics that are too live or too dull are both undesirable. In preschool facilities, offices, canteens or circulation areas without acoustic treatment, an unacceptably high noise level can build up at times.

But how do we solve the aforementioned, undoubtedly very diverse, acoustic problems? Should we view acoustics as an art rather than a science? In the end it is certainly a combination of both. Our current knowledge of the physical relationships and their subjective effects enable us to define room acoustics qualities and demonstrate through prognoses or calculations how these can be achieved with constructional measures. However, our acoustic plans must always be accompanied by intuition, experience and creativity. For ultimately it is not solely the objectively quantifiable sound propagation that determines the hearing experience; it is also influenced by design, colours and well-being. The aim of room acoustics design should therefore be to incorporate the room acoustics issues in the design of the interior space and its surfaces in a sensible way.

With that in mind, this part of the book is intended to provide an overview of the laws that apply to the propagation of sound in an enclosed space, how sound is perceived and, last but not least, how the acoustic quality of a room is determined by its shape and surfaces.

Sound propagation and auditive perception in rooms

It is not the room that determines the acoustics, but rather the sound source. However, the room does determine how the sound arrives at the listener. If a room consists exclusively of hard, smooth surfaces, the listener's impression is comparable to the visual impression gained in a room full of mirrors: too bright (= not understandable), too glaring (= too loud), and there is a lack of orientation. The reason for this is the sound bouncing off hard surfaces, just like light is reflected by very bright surfaces. By contrast, surfaces capable of "vibration" and open-pore surfaces absorb sound like darker colours absorb light.

But in contrast to light, the time component – i.e. when the reflections reach the ear – plays a crucial role with sound. This is easiest to understand if we emit a brief impulse, i.e. a bang, and track the propagation in a model by way of soundwaves (ray tracing, Fig. 1a).

The so-called direct sound reaches the listener first because it has to cover the shortest distance. It is followed by reflections from the ceiling and walls. Just like light, sound is reflected from flat surfaces such that the angle of incidence is equal to the angle of reflection. The interaural time differences are determined by the paths the soundwaves have to travel in the room.

The number of reflections increases with time, but the energy of the individual reflections decreases owing to the spherical sound propagation (p. 11, Fig. 2) and the losses due to absorption during the reflection process.

Fig. 1b shows such a schematic echogram – also known as echograph or room impulse response. When we speak or play music in a room, the signals are superimposed on each other in the same

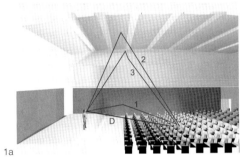
1a

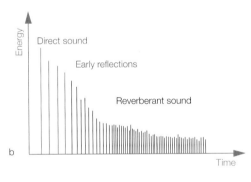
b

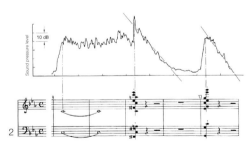
2

1 The sound propagation paths in a room can be illustrated with soundwaves.
  a Paths of direct sound and first reflections
  b Schematic room impulse response: direct sound and reflections shown by way of individual impulses.
2 The reverberation after fortissimo passages, like here in Beethoven's Coriolanus Overture, Op. 62 (bars 9 – 13), is readily discernible in the recording of the levels.

way and reach the ears of the listener thousands of times with corresponding time delays and weakening.
The individual reflections are not resolved by the ear; instead, together with the direct sound they determine the aural impression of a sound as a whole.

Direct sound
We use the direct sound to localise the sound source, i.e. it enables us to position a source in space even with our eyes closed – and we are able to do this even though the energy of all reflections together is considerably greater than that of the direct sound alone. If there are obstructions between the listener and the sound source, the direct sound can be weakened to such an extent that localisation is impaired. Guaranteeing an unobstructed direct sound propagation is therefore always important when acoustic intelligibility and clarity are important.

Early reflections
Reflections that reach the listener within 50 ms of the direct sound increase the intelligibility of speech owing to the ability of the ear to integrate those sounds. An interaural time difference of 50 ms corresponds to an approx. 17 m difference in the lengths of the paths travelled by the direct sound and the reflection. The intelligibility of music is further enhanced by reflections with an interaural time difference of up to about 80 ms (= 27 m difference in the paths travelled). Intelligibility means the distinguishability of successive reflections in a musical performance in a closed room despite superimposed diffuse sound.

Based on these fundamental relationships, it is possible to derive direct consequences for the room geometry and, in particular, for the line of the ceiling in larger venues. Such rooms should be designed so that early reflections are directed towards the listeners.
In addition, if the early reflections reach the ears of the listeners from the sides, this enhances the three-dimensional acoustic impression. This feeling of being "surrounded" by the music is these days an important quality criterion for concert halls in which symphony orchestras perform.

Reverberation
The early reflections are followed by the reverberant sound in which the density of the reflections increases and in many rooms the energy decreases at an approximately exponential rate. The reverberation of a room is the most important acoustic quality feature, especially since the reverberation, in contrast to the early reflections, is usually not, at best only marginally, dependent on position.

Which reverberation time is desirable for which room depends entirely on the function of that room. In cathedrals and churches, for example, a long reverberation time reinforces the sacred character and provides organ and choral works with the proper acoustic environment. In contrast to this, the reverberation time in lecture theatres should not be too long in order to avoid successive syllables being lost in the reverberations (although it is possible to adjust for this by speaking slowly).

Disturbing reflections
If high-energy reflections occur in the reverberant sound, these may be perceived as echoes, i.e. we hear the sound signal twice. Flutter echoes are periodically recurring reflection sequences which, for example, can build up between parallel wall surfaces. Such echo effects can disrupt music and speech quite considerably and should therefore be avoided.

Objective acoustic quality criteria
The above qualitative relationships between the structure of reflections and subjectively perceived listening attributes are found in objective room acoustics criteria. These can be measured in finished rooms or halls, or simulated or calculated in advance during the design work.

T1: Subjective effect of individual reflections or series of reflections

Time difference	Path difference	Subjective effect
≤ 1 ms	≤ 0.3 m	Sound colouration, possibly disruption to the sound source localisation
1 ms–50 ms	0.3 m–17 m	Increase in the intelligibility of speech due to the subjective "amplification" of the direct sound
1 ms–80 ms	0.3 m–27 m	Increase in the intelligibility of music if reflections come from the sides; enhanced acoustic three-dimensional impression
≥ 50 or 80 ms	≥ 17 or 27 m	Echo, i.e. hearing the sound signal twice
Periodic reflections at regular intervals		Flutter echo: hearing the sound signal several times; "buzzing" sound impression with path differences ≤ 17 m

1 Overview of reverberation times depending on
   room function. Which reverberation time is ideal in
   each individual case also depends on the volume
   of the room.

Reverberation time

By far the most important room acoustics criterion is without doubt the reverberation time, which describes the decay of the reflections over time. The reverberation time is the length of time taken for the sound energy to fall to one-millionth of the original level after switching off the sound source. Expressed as a level, this corresponds to a decrease of 60 dB.

One important initial step in room acoustics design is to specify the desired reverberation time depending on the use and volume of the room. Fig. 1 provides an overview. The reverberation time should not depend too much on the frequency, especially for the octaves from 250 to 2000 Hz. Deviations in the region of ±20% do not generally cause problems. But if the reverberation time for low frequencies is considerably longer than that for medium and high frequencies, the room will sound "boomy" and muffled, whereas extra

emphasis on the medium and high frequencies will lead to a bright to shrill aural impression.

In large concert halls, an increase in the reverberation time for low frequencies below the 250 Hz octave is desirable if a warm sound is to be achieved. In rooms for speech, this should be avoided because it impairs legibility.

Further acoustic quality criteria

Besides the reverberation time, there are many more objective room acoustics criteria, which are given more attention when larger venues are involved especially. The strength G is a relative measure of the total energy contained in the direct sound plus all reflections. It specifies how the source is "amplified" by the room and is therefore a measure of the relative loudness of a room.

The clarity C_{80} describes the acoustic legibility or intelligibility of music. It is the ratio of the sound energy incident within

the first 80 ms to the energy of the subsequent reflections.

A similar criterion, but used to assess the suitability of a room for speech, is the definition D.

Somewhat more meaningful is the speech transmission index (STI), which in a slightly different form can also be used for assessing SR systems and alarm installations.

The IACC (interaural cross-correlation coefficient) is used to describe the spaciousness of a room subjectively. This coefficient is determined by evaluating the similarity of the sound signals arriving at both ears. If the sound comes more from the sides, there is less similarity, which leads to a more spatial timbre, subjectively, and hence a low IACC. The IACC is especially popular for assessing concert halls for classical music.

The exact definitions of these room acoustics criteria, which are mainly used for the acoustic evaluation of larger venues, can be found in specialist publications and also in ISO 3382.

Relationship between volume of room, reverberation time and absorptive surface area

Owing to the sound-absorbing properties of clothes, rooms must always have an adequate volume in order to be able to achieve a certain reverberation time. Tab. T2 lists the room volume per seat, which can be used as an elementary acoustic requirement. If a figure is much lower than that given here, it is generally impossible to achieve very good objective acoustic characteristics for the type of use concerned. With higher values, more absorbent surfaces are required in order to reduce the reverberation time to a reasonable level. At the same time, this leads to a reduction in the loudness – normally an undesirable effect in rooms used for non-amplified music or speech.

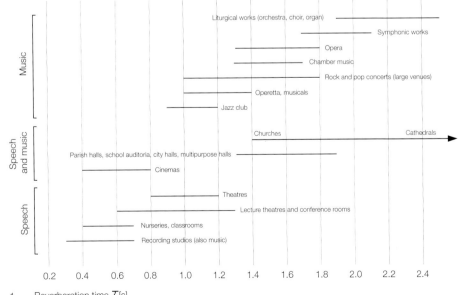

1   Reverberation time T [s]

T2: Acoustically favourable room volumes per seat

Room type/usage	Room volume per seat
Lecture theatre	4–6 m³
Theatre	4–6 m³
Multipurpose venue (music and speech)	6–9 m³
Symphonic music	10–11 m³
Churches	> 10 m³
Music rehearsal rooms	15–50 m³

Figures much lower than those given here mean that it is very difficult to achieve optimum reverberation.

These considerations become irrelevant when the sound propagation is exclusively by way of electroacoustic SR systems, e.g. large lecture theatres and congress halls, or arenas for rock and pop music or sports events. In such situations, even much higher volume/seat figures than those listed here are not critical, although the total area of absorbent surfaces required is larger.

More accurate, quantitative predictions of the reverberation time are possible when the volume of the room is known and the absorption of the room surfaces can be calculated using the so-called Sabine reverberation equation.

This, the most important equation in practical room acoustics, is named after the American physicist Walter Clemence Sabine, who around 1900 determined the reverberation time in a Boston lecture theatre using an organ pipe and a stopwatch. A few years later, Sabine was involved in the building of the Boston Symphony Concert Hall, world-famous today for its acoustics. Sabine is regarded as the father of modern room acoustics.

His reverberation equation is as follows:

$$T = 0.163 \cdot V/A,$$

where
T = reverberation time in s
V = volume of room in m³

A = so-called equivalent absorption surface area in m².

The relationship states that the reverberation time increases with the volume of the room, regardless of the room's geometry. The absorption surface area A is made up of the room surfaces S multiplied by their sound absorption coefficients α:

$$A = \alpha_1 \cdot S_1 + \alpha_2 \cdot S_2 + \alpha_3 \cdot S_3 + ...$$

The sound absorption coefficient α describes the ratio of the non-reflected sound energy to the incident sound energy and lies between 0 (fully reflected) and 1 (fully absorbed). An open window has a sound absorption coefficient of 1 because the entire sound energy leaves the room. Plastered wall surfaces, glazing and hard floor coverings exhibit sound absorption coefficients in the region of $\alpha = 0.04$ to 0.08. Seated persons (in rows of seats) exhibit a high sound absorption coefficient amounting to about 0.8.

The reverberation equation shows that a theoretical estimate of the reverberation times can be comparatively simple, provided the sound absorption coefficients of the surfaces are known. The sound absorption of architectural surfaces is investigated in more detail later in this section.

The difficulties in acoustic design tend to be of a more practical nature and relate to the fact that materials and surfaces in interiors are not chosen solely because of their degree of absorption, but rather according to architectural concepts, cost, mechanical durability, fire resistance, etc. The sound-absorbing effect of a surface, and hence also the reverberation time, depends on frequency, which is why calculations are normally carried out for the octaves from 125 to 4000 Hz. At high frequencies, the absorption effect of the air is taken into account as well.

In order to reduce the effect of frequency in the reverberation time calculation, consistent absorption across all frequency bands must be guaranteed.

The use of the Sabine reverberation equation presumes a "diffuse sound field". What this means is that the room is evenly "illuminated" acoustically and the incident sound reflections are evenly distributed in all directions. These assumptions apply to many interior spaces and types of room, at least approximately. But there are also exceptions such as geometrically simple rooms with asymmetric absorption (e.g. gymnasium with an absorbent ceiling only) or rooms with extreme acoustic attenuation (e.g. cinema, recording studio). And in rooms in which one dimension is considerably different to the others, e.g. open-plan offices, the Sabine reverberation equation supplies unreliable results.

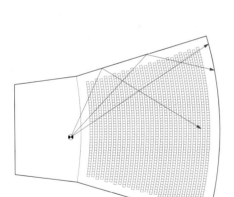

1a

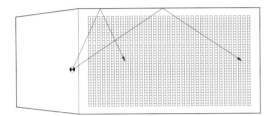

b

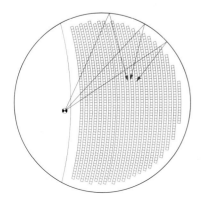

c

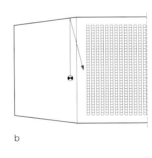

2a

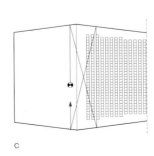

b

c

Loudness of a room

Once the conditions for a diffuse sound field are fulfilled, it is also possible to calculate the "loudness" of a room using a simple equation:

$$L_p = L_W + 10 \cdot \log[4/A]$$

This means that the sound pressure level depends solely on the sound power level of the source L_W (in dB) (p. 11) and the equivalent absorption surface are A (in m²). If we include direct sound as well, the equation is as follows:

$$L_p = L_W + 10 \cdot \log[1/(4\pi r^2) + 4/A],$$

where
r = distance from source in m.
This relationship shows that the loudness level in a room remains constant beyond a certain distance and depends only on the total absorption surface area A.

1 How the plan shape of a room affects the early reflections from the side walls
   a Narrow, rectangular rooms send reflections from the side walls to the seats.
   b If the room widens towards the rear, the sound is reflected to the rear of the room.
   c Concave plan shapes focus the soundwaves, which usually leads to disturbing concentrations of sound.
2 a The soundwave illustrates flutter echoes between parallel wall surfaces.
   b This can be avoided by designing the stage to taper towards the rear.
   c If the walls taper inwards a little, the soundwave is reflected back and unpleasant flutter echoes can be the result.

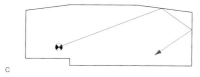

3a

c

3 a Reflections from the rear wall with a long delay
     increase the risk of echoes. Right-angles are
     particularly treacherous because the sound is
     reflected back parallel to its original direction.
     Disturbing reflections can be avoided by...
   b including absorbent surfaces,
   c positioning the rear part of the ceiling at an
     angle,
   d including structured wall surfaces.

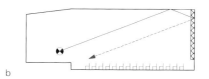

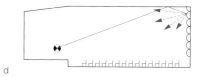

b

d

The distance beyond which the energy of the reflections exceeds that of the direct sound statistically is known as the diffuse field distance r_H:

$$r_H \times 0.06 \cdot (V/T)^{1/2} \times 0.14 \cdot A^{-1/2}$$

where r_H is in m, T in s, V in m³ and A in m².

In a classroom with a volume of 200 m³ and a reverberation time of 0.7 s, at a distance of just 1 m the diffuse sound, i.e. the total of all reflections, is already dominant.
By doubling the absorption surface area in a room, which is equivalent to halving the reverberation time, the sound pressure level outside the diffuse field distance can be reduced by 3 dB. Furthermore, doubling the room volume while the reverberation time remains equal or similar results in a 3 dB reduction in the loudness level. These are just some of the considerations necessary when designing for noise control in noisy workplaces.

Room geometry and reflections structure

Ray geometry methods are popular when we want to move away from statistical considerations and assess and demonstrate the influence of room shape and orientation of the surfaces. Figs. 1–3 illustrate a number of fundamental relationships.

Structured surfaces

Specular reflections (angle of incidence = angle of reflection) occur only when surfaces, or parts thereof, are sufficiently large and smooth. Surfaces are acoustically smooth (equivalent to a shiny surface in optics) when the depth of any surface structure is less than about 1/12 of the wavelength. For example, surfaces with a 30 mm deep structure reflect

sounds < 1000 Hz geometrically (wavelength for 1000 Hz = 0.34 m).
On surfaces with a deeper structure, the sound is reflected not in one direction, but rather scattered over a wide solid angle. A depth of at least 200 mm is necessary when lower frequencies are to be scattered as well. The surface structure of woodchip wallpaper is certainly not adequate for this.
Fig. 3 (p. 19) shows schematically the three-dimensional breakup of the sound reflections at a structured surface.

Surface structures can also be used to improve the acoustics of difficult room geometries with concave curving surfaces or parallel walls (p. 18/19, Fig. 1). Moreover, structured surfaces are particularly important in rooms where good acoustics are vital, e.g. concert halls, rehearsal rooms, recording studios. And such surfaces can certainly be employed as interior design elements. In older concert halls, columns, pier shafts and ornamentation make a decisive contribution to good acoustics.
Resolving larger surfaces into smaller, possibly convex curving, surfaces or providing three-dimensional projections such as window and door reveals or post-and-rail constructions can also help to improve the sound mix.
Scattering of the sound at irregular surfaces is therefore a highly desirable effect because this defocuses the reflections and causes a better sound mix.
But this does not reduce reverberation noticeably. With a reflective material (wood, stone, glass), a structured surface does not increase the sound absorption significantly – only roughly equivalent to the extent to which the surface development has been enlarged.
Pleasing from the acoustical viewpoint is that structured surfaces are becoming popular again in architecture.

More accurate prognosis methods and room acoustics measurements

Computer simulations for room acoustics
These days, computerised acoustic simulations offer additional planning dependability and occasionally enable designs to enter "acoustic grey zones".
Computer simulations studying the acoustics of interiors mostly employ geometrical techniques such as ray tracing and the mirror source method.
The architect's drawings form the basis for a three-dimensional model of the interior, which is restricted to data relevant to the acoustics (p. 19, Fig. 2a). The surfaces are assigned frequency-related sound absorption coefficients to suit the intended materials (publications, manufacturers' data, empirical values). Any surface structures are taken into account by way of a degree of scatter. The degree of scatter describes the ratio of scattered energy to total reflected energy and is usually defined empirically, but has recently also been measured in laboratory tests.
Several thousand soundwaves are emitted simultaneously from defined transmitter positions and their paths through the room are traced. Once a soundwave strikes an enclosing surface, its energy is reduced depending on the absorption coefficient. A flat, smooth surface reflects the soundwave geometrically. With other surfaces, the direction of reflection is defined taking into account a scatter characteristic.
The result of such simulations are room impulse responses for typical receiving positions, or rather seating positions, which can be further analysed with respect to their reflection distribution. It is also possible to estimate and assess the aforementioned room acoustics criteria for complete auditoria (Fig. 2).

The accuracy and significance of the results depend quite decisively on the

1a   b

choice of input parameters, i.e. modelling of the surfaces, absorption and scatter properties. If the modelling contradicts the laws of geometric acoustics, e.g. the surfaces are resolved too finely, unrealistic results are the inevitable outcome. Used with care and backed up by experience, room acoustics simulations are a valuable aid in modern acoustic design. With a little more effort it is possible to use the results of the simulation to hear what rooms sound like in advance. The term auralisation was coined for this. In this case the receiving characteristic of the human ear is emulated in the simulation. It is possible to create a realistic aural impression of a room by processing speech or music recordings (recorded without any reverberation) and played back via headphones or loudspeakers. This is a very good way of assessing speech or loudspeaker systems, or also individual instruments. However, emulating the sound of a whole orchestra is still a dream owing to the complexity and interaction of the sounds.

Special features of small rooms
Describing the sound propagation with the help of geometrical and statistical methods is inadequate for small rooms with a volume less than approx. 100 m³ and low frequencies less than about 160 Hz. The wavelengths are then in the order of magnitude of the room dimensions and the sound pressure level depends on how good the respective wavelength "fits in the room". Room resonances occur when one dimension of the room coincides with half the wavelength or a multiple of it. Undamped room resonances make themselves felt during speech or music as unpleasant booming. This is particularly noticeable when there is a whole-number ratio between length, width and height because then the same resonant frequencies are superimposed.

Favourable proportions for rectangular rooms require room dimensions that are as dissimilar as possible, e.g. ratios of 1:0.83:0.47 or 1:0.79:0.62. Placing one or more room surfaces at an angle, dividing up large areas of the wall surfaces or using absorbent materials are suitable methods for suppressing disturbing room resonances. Such aspects are particularly important for the recording and listening rooms of studios, but also in classrooms for music. These considerations also play a role in small, possibly glazed, offices. The physical effects can be predicted with more accuracy by using finite element or boundary element methods, for example, where the sound is modelled according to its wave nature. For large rooms, such methods are only suitable for low frequencies at best owing to the computing time required.

Room acoustics measurements
An objective room acoustics quality assessment, or a report on the current situation, is necessary to check both design and construction, in advance of planned refurbishment work, or in the event of complaints. Whereas in the past a blank cartridge fired from a gun or bursting balloons were the methods often used, these days synthetic measurement signals that can cover the entire range of audible frequencies are used. In this way, room impulse responses can be determined quickly and accurately, allowing the reflections structure and objective room acoustics criteria to be evaluated. It is also possible to track down acoustic defects with the help of intelligent measuring techniques.
If only the frequency-related reverberation time is required, the decay process of noise signals after switching off can be evaluated.

Measurements of room acoustics are generally carried out with the help of spe-

cial measuring loudspeakers. Microphones or possibly a dummy head (p. 11) are used as receivers. The number of measuring positions (transmitters and receivers) varies with the size of the room and lies between about six positions in rooms of classroom size up to more than 100 in concert halls and opera houses. The duration of the measurements varies correspondingly, from 30 minutes to whole days (or nights).
The measurements are mostly carried out in unoccupied rooms and the results converted to the occupied condition. Measurements in venues are occasionally carried out in the occupied condition, e.g. directly prior to a performance, with fewer measuring points (duration about 5–10 min).

Measurements on models
During the design phase it is possible to measure the room acoustics using models, normally built to a scale of 1:10 or 1:20. The model includes all the surfaces in the room shaped and positioned according to the drawings. Owing to the smaller dimensions, the lengths of the soundwaves are also scaled down, i.e. the measurements are carried out at higher frequencies.
The problems are that absorption increases with the frequency and this affects the measurements for sound propagation in air, and it is not easy to transfer the sound absorption properties of materials to the scale of the model. The audience, or seating, is usually represented by sound-absorbent profiled plastic foam (Fig. 2b).

Compared to computer simulations, measurements on models have the advantage that the sound propagation remains faithful to the wave nature of sound, i.e. focusing and scattering of the sound is emulated properly in physical terms.

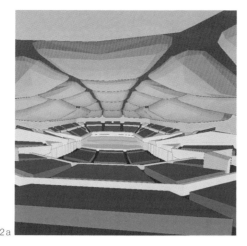

2a

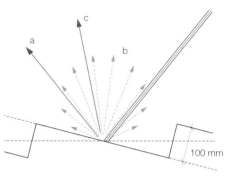

3

100 mm

Owing to the different pros and cons of measurements on models and computer simulations, both methods are sometimes used together for particularly demanding room acoustics tasks (concert halls, opera houses).

Sound absorption of architectural surfaces

Many of the surfaces preferred by architects these days, e.g. glass, fair-face concrete, plaster, reflect sound. It is therefore frequently necessary to incorporate sound-absorbent materials into the interior design. Industrially manufactured products, custom-designed surfaces and multi-layer constructions are all potential options.

All kinds of sound-absorbent systems are available, from mineral-fibre insulating boards, laid, for example, in suspended ceiling systems, to special products such as microperforated foils. The materials differ not only in terms of appearance and price, but also in their acoustic efficiency, i.e. the sound absorption coefficient.

Measuring sound absorption in a reverberation room
Sound-absorbent linings and materials are identified by the sound absorption coefficient for omnidirectional sound incidence. Typical values for many architectural surfaces can be found in the relevant publications. For acoustic products, the sound absorption coefficients are listed in their technical specifications. Most of the data is based on measurements carried out in a testing laboratory to DIN EN ISO 354. According to this international standard, the construction to be tested – an area of 10–12 m^2 – is set up properly in a so-called reverberation room (Fig. 1a). After measuring the frequency-related reverberation times with and without the construction, the Sabine

1 Examples of wall linings that scatter the sound:
  a Bruneck Grammar School, Southern Tyrol
  b Refurbishment of Teatro Reggio, Turin
2 Sala Santa Cecilia, Parco della Musica, Rome, 2002, Renzo Piano Building Workshop
  a Computer model for acoustic purposes: the room acoustics computer simulations help to provide more accurate forecasts of the acoustic propagation quality that can be expected.
  b Model built for acoustic tests (scale 1:20), shown here opened; the audience is modelled by soft profiled acoustic foam
3 Soundwaves are reflected in various directions from structured surfaces; schematic drawing of a lining with a sawtooth structure.
  a Low frequencies ignore the structure if the wavelength is large compared to the dimensions of the structure; the sound energy is reflected geometrically with respect to the dotted red line.
  b Medium frequencies are scattered more or less evenly in different directions.
  c High frequencies are reflected geometrically from the individual surfaces (dotted blue line) because here the dimensions of the sawtooth structure are large compared to the wavelengths.

2b

1a

1   a The measurement of the sound absorption of
       architectural surfaces is carried out in a rever-
       beration room according to ISO 354 using
       sample constructions measuring 10–12 m². The
       reflective, curved panels suspended from the
       ceiling ensure the correct sound mix.
    b The result of such measurements is the fre-
       quency-related sound absorption coefficient,
       which is recorded in a test log.

reverberation equation can be used to determine the sound absorption coefficient α_s, which is specified in one-third octave bands from 100 to 5000 Hz. Practical sound absorption coefficients α_p in octave bands can then be summarised from the results of the tests according to DIN EN 11654 "Acoustics – Sound absorbers for use in buildings – Rating of sound absorption" (July 1997). These values are also listed in specifications in order to define the frequency-related sound absorption of linings or ceiling systems.
A single value, the weighted sound absorption coefficient α_w, is determined from the practical sound absorption coefficient, which comprises six numerical values for the octaves from 125 to 4000 Hz. The weighted sound absorption coefficient can be entered into initial calculations that provide a rough orientation and enable different products to be compared. In tenders for absorbent linings, it can be used together with a description of the construction to specify the acoustic performance required. However, the weighted sound absorption coefficient is unsuitable for more accurate acoustic calculations.

Measuring sound absorption in an impedance tube
A different measuring method is sometimes used in the course of initial estimates or product development. This method requires substantially smaller sample dimensions – typically 100 mm diameter – and involves measuring the sound absorption coefficient α_0 for perpendicular sound incidence in a so-called impedance tube. But this method is less suitable for direct room acoustics applications owing to the exclusively perpendicular incidence of the sound. In interiors the sound strikes surfaces from different directions, which is why the results from the reverberation room are better.

Sound absorption mechanisms
Good sound absorbers are open-pore structures such as fibrous insulating materials, open-cell plastic foams, textiles, granulates or aerated concrete. Critical for the absorbing effect is that sound can infiltrate the pores, where the sound energy, which is nothing more than rapidly oscillating air, can be eliminated by friction.
The porosity, i.e. the ratio of the accessible volume of air to the total volume of material, of good sound absorbers exceeds 50%; indeed, mineral-fibre insulating materials consist of > 98% air.

In contrast to this, closed-cell plastic foams, e.g. rigid polystyrene foam, are just as unsuitable as sound absorbers as water, which although soft, does not allow the sound to infiltrate better than concrete. The most important physical property alongside porosity is the acoustic airflow resistance, which describes the resistance the sound has to overcome. In contrast to the static air permeability of textiles, the acoustic airflow resistance is a dynamic variable. It can be determined according to DIN EN 29053 in the laboratory using small samples (approx. 200 × 200 mm), allowing, for example, the acoustic suitability of curtains to be assessed.
In addition, the thickness and positioning of an absorber influences the absorption considerably. This is illustrated in Fig. 2 (p. 22) by means of the sound absorption coefficient curves for three materials: a carpet covering, a 15 mm acoustic plaster (applied directly to a hard surface) and a 30 mm mineral-fibre insulating material.
It can be seen from the diagrams that the absorption is low at low frequencies. The rise in the absorption with the frequency increases even more as the thickness of the material increases because the friction

losses and hence the sound absorption are greatest when the velocity of the air particles in the material are highest. The movement of the air particles drops to zero as they are stopped directly at the hard wall. When the incidence of the sound is perpendicular, the velocity is a maximum at a distance of one-quarter of the wavelength (approx. 0.85 m for 100 Hz, approx. 0.34 m for 250 Hz and approx. 85 mm for 1000 Hz).

Owing to the omnidirectional incidence of sound in rooms, it is not necessary to fill an entire void with absorbent material. Instead, it is generally sufficient to provide thinner layers. For example, a ceiling of 20 mm thick absorbent panels suspended 200 mm below the structural floor increases the absorption effect considerably for lower frequencies (p. 22, Fig. 2).
These relationships show why simply laying a carpet or applying "acoustic wallpaper" have only a limited effect. As both materials are very thin and are positioned directly in front of a reflective surface, reverberation is reduced for high frequencies only.
Perforated or slotted boards made from plasterboard, wood-based products or metal, grid structures or woven metal meshes are frequently used, with a backing of sound-absorbent material. The absorptive effect depends on the proportion of openings, but also on the size of the openings and the thickness of the material (p. 22, Fig. 1). In qualitative terms we can say that the larger the proportion of openings, the greater is the sound absorption at higher frequencies. Typical proportions lie in the order of magnitude of 15%. Small openings are better than large ones, for the same proportion of openings. Typical hole diameters or slot widths lie between 1 and 20 mm. In order to achieve an adequate absorptive effect, also in the high-frequency

range, the pitch of the openings should normally be less than about 100 mm. Thin boards or sheet metal allow the passage of more sound than thicker boards, for the same proportion of openings. Wood-based board products exploit this principle by having a small proportion of openings on the visible side but a larger proportion on the back. Such boards still possess sufficient mechanical stability and at the same time the acoustically effective thickness of the board is reduced. Micro-perforated or very finely perforated plywood – bonded to a sound-absorbing backing board – owe their absorptive effect over a wide range of frequencies to the fact that the perforated layer is very thin.

Plain boards can achieve significant sound absorption, especially in the low-frequency range. One example is a wooden lining that is mounted clear of the wall; in terms of physics this is a mass-spring system. The mass of the lining per unit area and the compressibility of the air determine the resonant frequency at which maximum absorption occurs:

$$f = \frac{540}{\sqrt{m' \cdot d}}$$

(f in Hz, m' in kg/m^2, d in cm).

When the mass per unit area of a wooden lining is in the order of magnitude of 6 kg/m^2 (8 mm plywood) and the clearance to the wall is 60 mm (air + damping), this results in a resonant frequency of approx. 90 Hz. The lighter the board, the higher is the absorption at this resonance. In addition, the damping of the void plays an important role.

b

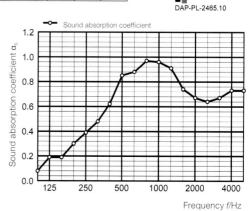

Sound absorption coefficient to ISO 354
Measurement of sound absorption in a reverberation room

Client:
Test specimen: Acoustic panels made from glass granulate

(from top to bottom):

- approx. 18 mm Acoustic panels made from glass granulate
  Panel front face: decorative acoustic coating
  Panel rear face: sound-absorbing covering
  6 pieces laid loose, butted together, butt joints masked
- 50 mm Cavity without damping, with supporting construction
- Floor of reverberation room

Enclosing frame made from 19 mm coated particleboard
Joints between frame and floor of reverberation room and between frame and panels masked

Room: reverberation chamber E
Volume: 199.60 m^3
Area of test specimen: 10.00 m^2
Date of test: 6 June 2007

Testing laboratory accredited
to ISO/IEC 17025
Akkreditiertes Prüflaboratorium
nach ISO/IEC 17025

Deutscher
Akkreditierungs
Rat

DAP-PL-2465.10

	Θ [°C]	r.h. [%]	B [kPa]
w/o specimen	21.4	45	95.5
with specimen	21.2	46	95.6

Frequency [Hz]	α_s one-third octave	α_p octave
100	° 0.08	
125	0.19	0.15
160	0.19	
200	0.30	
250	0.39	0.40
315	0.48	
400	0.62	
500	0.85	0.80
630	0.88	
800	0.97	
1000	0.96	0.95
1250	0.91	
1600	0.74	
2000	0.67	0.70
2500	0.64	
3150	0.67	
4000	0.73	0.70
5000	0.73	

° Absorption surface area < 1.0 m^2
α_s Sound absorption coefficient to ISO 354
α_p Practical sound absorption coefficient to ISO 11654

Sound absorption coefficient

Sound absorption coefficient α_s

Frequency f/Hz

Rating according to ISO 11654:

Weighted sound absorption coefficient αw = 0.70 (M)
Sound absorption class: C

MÜLLER-BBM

Planegg, 18 May 2007
Test report No. M69 749/8

Annex A
Page 1 of 3

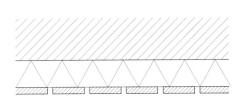

1a

○—○ 50 mm fibrous insulating material with fleece facing
●—● with 60 mm battens and 60 mm joints
□—□ with 60 mm battens and 10 mm joints
■—■ with plain 8 mm plywood

b

Ceiling and wall systems

The most diverse absorbent materials are available, such as fleece-backed mineral insulating boards for lay-in ceiling systems at prices from about 25 €/m², perforated plasterboard for about 60 €/m², and specially slotted and perforated linings made from wood-based products or acoustic plaster systems on backing boards, which quickly reach prices in excess of 150 €/m². The weighted sound absorption coefficients for the typical construction depths of between 70 and 200 mm lie between about α_w = 0.55 and 1, depending on the particular product.

Sound-absorbent linings can also be considered for the walls. Aspects such as mechanical stability, susceptibility to soiling and hence cleaning and renovation requirements along the lower part of the wall must be considered when selecting materials and systems.

When using perforated or slotted surfaces on the walls, the risk when using lighter colours in conjunction with dark holes (fleece backing) is that the surface appears to flicker when viewed close-up. In cases of doubt, a sample should be applied to a larger area first to test the effect.

Further possibilities for walls are masonry or concrete absorbers, e.g. perforated acoustic clay bricks, special concrete blocks or aerated concrete surfaces.

Backings

The sound absorption of many types of product can be increased considerably if an absorbent backing is applied over the full area. Boards or sheets made from mineral-fibre insulating materials (DIN 4102 combustibility class A) can be used, wrapped in very thin PE sheeting to prevent fibres escaping into the interior air. Boards made from polyester fleece (B1) or melamine foam (B1, A) are also possible as well as blankets made from sheep's wool or cotton (fire resistance dependent on chemical additives). Backings are typically between 20 and 50 mm thick.

Absorber elements

Ceiling panels, but also absorbent demountable partitions or furniture can be classed as absorber elements, for example. Decorative acoustic panels represent another possibility; they can be used as notice boards and are being used more and more often.

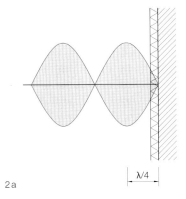

2a

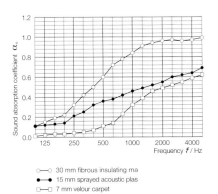

○—○ 30 mm fibrous insulating ma
●—● 15 mm sprayed acoustic plas
□—□ 7 mm velour carpet

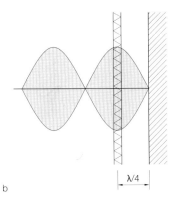

b

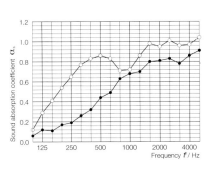

○—○ 20 mm acoustic foam with 200 mm clearance to reflective surface
●—● 20 mm acoustic foam attached directly to reflective surface

1 a, b The sound absorption of perforated or slotted boards depends on the proportion of openings. The smaller the proportion, the greater the absorption is diminished at high frequencies. Plain boards exhibit a significant absorption at low frequencies.
2 How the positioning of the absorbent material in front of the reflective surface affects the outcome:
   a The movement of the air particles is equal to zero directly at the wall and the absorption due to friction in a thin absorber is low. For this reason, thin materials absorb sound only at high frequencies.
   b The velocity reaches a maximum at a distance of one-quarter of the wavelength. If this oscillation range is located within the absorber, maximum absorption is to be expected (schematic diagram for perpendicular sound incidence). Increasing the size of the void improves the sound absorption at lower frequencies considerably.

3 Sound-absorbent linings are available in the most
   diverse designs:
   a Finely perforated veneer on absorbent backing
      board made from glass granulate
   b Perforated/slotted wood-based board product
      with absorbent backing
   c Wood-wool acoustic board
   d Acoustic plaster system
   e Woven metal mesh with absorbent backing

With absorber elements, the absorption effect per unit area is greater than large-scale arrangements, partly because the sound at the edges is "sucked" into the material. This can lead to sound absorption coefficients much higher than 1, which appears nonsense in terms of physics. In order to avoid misunderstandings, absorber elements are therefore defined via an equivalent sound absorption surface area A per object, which is likewise measured in a reverberation room.

Curtains
Curtains or sheets of fabric can contribute noticeably to the attenuation in a room and are a popular way of varying the acoustics in, for example, music rehearsal or multipurpose rooms.

To achieve a high degree of the absorption of $\alpha_w = 0.6$ or more with curtains calls for a weight per unit area of at least approx. 300 g/m^2 and a specific airflow resistance in the region of $R_s = 800–2500$ Pa s/m.

It is possible to gain an impression of the acoustic suitability by blowing through the material. If this is easy, e.g. with a transparent gauze material, the acoustic resistance is low and little absorptive effect can be expected. But if it is impossible or very difficult to blow through the material, the material is obviously too dense, which leads to the sound tending to be reflected rather than absorbed. The test according to DIN EN 29053 (see above) enables an objective assessment.

Absorbent flooring systems
As already described, laying carpet directly on a hard substrate only achieves good sound absorption for high pitches. The sound absorption can be extended to a much wider range of frequencies if a void below the carpet is made to act

acoustically. This requires the use of, for example, perforated raised access floor panels in conjunction with carpeting suitable for use with displacement ventilation (with suitable flow impedance). The absorptive effect depends on the acoustic quality of the carpet, the proportion of holes, the thickness of the backing panel and the depth of the void.

Seating
In venues and churches, upholstered, sound-absorbent seats or pews ensure a basic acoustic damping. Seating with good acoustic properties is able to minimise the difference in the acoustic conditions between unoccupied, low occupancy (e.g. rehearsal situation) and fully occupied rooms.

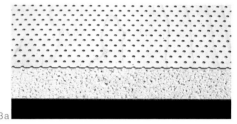

3a

b

c

d

e

Building acoustics

Building acoustics is concerned with the propagation of sound within a building, from room to room, from outside to inside, possibly even from inside to outside. In other words, it concerns the prevention of sound propagation in order to avoid unreasonable acoustic disturbance.
The basis for planning work and legislation in Germany is usually DIN 4109 "Sound insulation in buildings", which is part of building legislation. DIN 4109 lays down minimum sound insulation requirements for residential buildings, educational establishments, hospitals and offices. But even for those buildings not covered by the standard, e.g. detached houses, private offices, even cultural and conference centres, it is advisable to define a sound insulation standard and implement this in the building design.
It is against this background that sound propagation within a building and its characteristic variables are presented on the following pages. One aspect of this is how to establish the resultant sound insulation through "adding" the sound insulation of individual components. The relationships between the construction of components and their sound-insulating properties are also illustrated.
The requirements for sound insulation in buildings are not given here in terms of numbers. Instead, there are references to the chapters on residential, office and school buildings plus facilities for music. The chapter on residential buildings also includes more information on the changes to DIN 4109, a revised edition of which is due to appear shortly.

Sound propagation in buildings
Airborne sound excitation
Speaking in a room, for example, causes the enclosing components to vibrate. These oscillations propagate within the construction (so-called structure-borne sound) and are radiated in, for example, an adjacent room in the form of airborne sound. The sound propagation takes place not only via the separating component, but also – and frequently in a similar order of magnitude – via the adjoining, so-called flanking components (Fig. 1). Accordingly, building acoustics has to consider and evaluate both the separating and the flanking components.

Structure-borne sound excitation
Where walls or suspended floors are not excited by airborne sound, but instead are caused to vibrate by way of direct mechanical actions, we speak of structure-borne sound. This is particularly the case when walking across a floor (impact sound), or when moving chairs, but also when using sanitary facilities and operating building services. The sound transferred into components propagates through the construction as structure-borne sound and is radiated in neighbouring rooms in the form of (secondary) airborne sound.

Airborne and structure-borne sound excitations often occur together, e.g. when using lifts and ventilation systems.

Vibrations
Vibrations are generally low-frequency structure-borne sound excitations (below about 63 Hz) which, for example, are caused by trains, construction activities or industrial operations. If such vibrations could have negative effects for people, historical buildings or sensitive laboratory apparatus, dynamic analyses are usually required, an aspect that exceeds the scope of the relationships dealt with here.

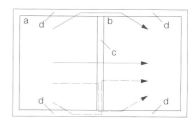

1

Noise control through planning of the layout
The foundation for good sound insulation can be laid right at the preliminary planning stage by designing the layout of a building so that noisy areas and areas needing protection from noise are kept apart.
For example, plant or boiler rooms should not be located adjacent to bedrooms, but instead below storage areas or corridors, for instance. In apartment blocks, bathrooms and kitchens should be positioned adjacent to each other, horizontally and vertically, wherever possible, which also favours the construction of service ducts. Another advantageous arrangement is to place a floor of offices between a restaurant at ground floor level and any apartments above. Another important consideration for the layout is noise from outside the building.
But even when the layout cannot be planned solely with respect to noise control, it can sometimes be helpful to be aware of such aspects.

1 Separating and flanking components and
   transmission paths (schematic)
   a Source room
   b Receiving room
   c Separating component
   d Flanking component

25

2

1 Frequency-related sound reduction index for a 200 mm thick reinforced concrete wall. The weighted sound reduction index R_w is determined by translating a reference curve in 1 dB steps until the sum of the negative deviations is just less than 32 dB. The value of the translated reference curve at 500 Hz then corresponds to the weighted sound reduction index – R_w = 58 dB in this example. This procedure may seem a little arbitrary, but it does approximate to the frequency-dependent sensitivity of the human ear and the typical behaviour of solid components, and has become accepted internationally (method according to ISO 717-1).

2 Standard tapping machine for measuring impact sound insulation

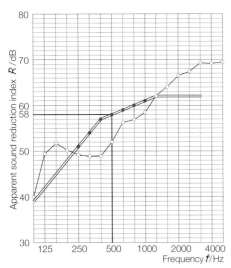

=== Shifted reference curve to ISO 717-1
—○— Apparent sound reduction index
• Negative deviation for shifted
1 ○ reference curve

Designating the sound insulation of components

Sound reduction index

The insulation of walls, suspended floors and doors against airborne sound is described by way of the sound reduction index R. This index specifies the number of decibels by which the sound is weakened as it passes through the component. The sound reduction index is therefore a component-related variable.

As the sound insulation of components depends on frequency, the sound reduction index is also specified depending on the frequency, at least in the one-third octave bands between 100 and 3150 Hz. For simplicity, a single value, the so-called weighted sound reduction index R_w, (designated by the suffix "w"), is derived from the frequency-related values. The calculation is carried out by comparing the frequency-related values with a standardised reference curve (idealised sound reduction index for a solid wall, see also Fig. 1).

Sound reduction indexes for many components and products, e.g. masonry walls, lightweight walls, glazing, also doors and movable room dividers, are available from their manufacturers. These are generally the results of tests carried out on typical constructions in the laboratory according to an international test method (ISO 140-3).

Flanking sound level difference

The flanking sound level difference D_{nf} (formerly the flanking sound reduction index R_L) is used for flanking components. This parameter describes the flanking transmissions of facades, hollow floors and suspended ceilings.

Further parameters

Besides the above variables for describing the airborne sound insulation, there are other definitions, e.g. the element sound level difference $D_{n,e,w}$ for designating the sound insulation of small components with an area < 1 m². The various terms can lead to confusion. To ensure an unambiguous specification of the sound-insulating properties of components, e.g. in building specifications, a clear description must be included in cases of doubt.

Weighted normalised impact sound level

The sound propagation caused by the structure-borne sound excitation of, for example, suspended floors or stairs when persons walk on them is described by the weighted normalised impact sound level $L_{n,w}$. Low levels correspond to better impact sound insulation.

The structure-borne sound excitation is simulated by a standard tapping machine (Fig. 2). The sound pressure level L_2 in the "receiving room" is measured while the standard tapping machine is operated in the room above or alongside.

In order to take into account the influence of the damping in the room, the sound pressure level L_2 is adjusted by the absorption surface area A_2 of the receiving room.

$$L_n = L_2 + 10 \cdot \log (A_2/10 \text{ m}^2).$$

where

L_n: normalised impact sound level
L_2: sound pressure level
A_2: equivalent absorption surface area in receiving room (in m²)

As for the sound reduction index, a single value for the weighted normalised impact sound level $L_{n,w}$ is determined from the frequency-related curve of the normalised impact sound level.

Reducing impact sound

Floating screeds or resilient floor coverings such as carpets can be laid on structural floors in order to reduce the impact sound transmissions. Such components and constructions are taken into account in acoustics by way of the weighted impact sound reduction index ΔL_w in dB. This parameter specifies the number of decibels by which the weighted normalised impact sound level of a solid structural floor is reduced by the floor finishes.

Designation of sound insulation

Sound reduction index R'_w
Variables identical or similar to those used for describing the sound insulation of components are used to designate the sound insulation of a construction. The critical difference here, though, is that the variables do not concern just the separating component, but instead take into account all transmission paths. In the end, it is irrelevant to your neighbour whether the sound arrives mainly via the separating wall or via a possibly acoustically less effective lightweight external wall! The dash (') suffix indicates that the total sound insulation, including flanking transmissions, is involved. For example, the airborne sound insulation between two rooms is designated by way of the weighted sound reduction index R'_w. This value includes the sound propagation via flanking paths, bypassing the separating wall, to a certain extent. The weighted sound reduction index with flanking paths R'_w is therefore always lower than the sound reduction index R_w of the separating component alone. If the propagation via the four flanking components (floors above and below, two walls) is on average similar to that via the separating component, which is certainly typical, the difference is a remarkable 3–5 dB. If the sound insulation via a flanking component is very low, e.g. in the case of a continuous floating screed, it is this component that determines the resultant sound reduction index.

Typical weighted sound reduction indexes are R'_w = 40 dB (office partition), R'_w = 53 dB (party wall between apartments) and R'_w = 67 dB (double-leaf party wall with high acoustic quality between terraced houses).

Sound pressure level in adjoining room
But how loud it is in an adjoining room does not just depend on the weighted sound reduction index of a component plus its flanking paths R'_w. The sound power passing through the construction increases with the area of the wall surface S. If the rooms are very reverberant, the numerous reflections mean that the loudness level is higher than in an attenuated room. But in contrast to room acoustics, building acoustics does not consider the exact structure of the reflections. Instead, we assume a "diffuse sound field", and the acoustic damping in the room is described solely by way of the total absorption surface area A (pp. 14–16). Using these relationships, the level in the adjoining room ("receiving room") L_2 can be estimated as follows:

$$L_2 = L_w - R' + 10 \cdot \log \left(\frac{4\,S}{A_1 \cdot A_2} \right)$$

where
$A1$ and $A2$ = absorption surface areas in source room and receiving room respectively
L_w = sound power level generated in source room
S = surface area of separating component

This relationship also shows that besides increasing the sound reduction index, absorption measures in the rooms can reduce the level in the receiving room. Doubling the absorption surface area A_1 or A_2 of a previously reverberant source or receiving room respectively reduces the received level L_2 by 3 dB in each case.
If the sound pressure level in the source room L_1 is known (e.g. by way of measurements), the above equation can be simplified to the following:

$$L_2 = L_1 - R' + 10 \cdot \log (S/A_2)$$

T1: Weighted sound reduction index terminology

R'_w	including flanking transmissions, describes standard of sound insulation
R_w	without flanking transmissions, applies to individual components, doors
$R_{w,P}$	without flanking transmissions, measured in the laboratory according to ISO 140-3
$R_{w,R}$	without flanking transmissions, characteristic value for verification of sound insulation to DIN 4109, corresponds to R_w $R_{w,R} = R_{w,P} - 2$ dB (walls, floors, windows) $R_{w,R} = R_{w,P} - 5$ dB (doors, movable room dividers)

The S/A_2 ratio is often in the region of 1, and then the sound reduction index more or less reflects the frequency-dependent difference in levels between the rooms.

Weighted standardised sound level difference $D_{nT,w}$
According to the revised edition of DIN 4109, in future the so-called weighted standardised sound level difference $D_{nT,w}$ will be used instead of R'_w. This corresponds better to the sound pressure level difference between two rooms and therefore better describes the true sound insulation:

$$L_2 = L_1 - D_{nT} + 10 \cdot \log (T/T_0)$$

where
T_0 = reference reverberation time (0.5 s for housing)
The weighted standardised sound level difference is related to the propagation with flanking paths and numerically is similar to the weighted sound reduction index R'_w.

Weighted normalised impact sound level $L'_{n,w}$
The impact sound insulation is designated by the weighted normalised impact sound level $L'_{n,w}$. Here again, the dash (') means that flanking transmissions are included. In the case of impact sound being transmitted to the rooms directly below, the

flanking transmissions are usually low. It is only when solid walls are considerably lighter than the structural floors that $L'_{n,w}$ is about 4 dB higher than $L_{n,w}$.
Typical weighted normalised impact sound levels lie between $L'_{n,w}$ = 78 dB (for solid structural floors without finishes) and 39 dB (for suspended floors with good-quality floating screeds), which corresponds to a good standard of sound insulation for apartment blocks.

Owing to the influence of body weight and footwear, it is not possible to specify a direct relationship between the normalised impact sound level and the level upon walking across a suspended floor. Furthermore, the impact sound level does not say anything about how loud walking sounds in the same room (see p. 33, "Suspended floors and floor finishes").

Analysis of sound insulation

In the end it is the designer who is responsible for ensuring that the sound insulation promised by legislation is actually achieved by the building. This can be checked during the design work with the help of a sound insulation analysis, which means the calculation of the theoretical sound insulation between individual rooms for the components of the intended design according to DIN 4109. The need to verify the sound insulation is handled differently in the different building codes of the German federal states. However, verification of sound insulation is not a substitute for careful planning of details with respect to their acoustic performance.
When checking the sound insulation, the following "playing rules" should be kept in mind (based on DIN 4109 supplement 1).

Tolerance allowance
The characteristic value for the sound reduction index can frequently be determined directly by means of tables (e.g.

DIN 4109 supplement 1). The input variable can be, for example, the mass per unit area of a masonry wall. When using sound reduction indexes derived from laboratory measurements, however, a safety factor has to be included. The background to this so-called tolerance allowance is that specimen constructions in the laboratory are sometimes built more carefully than is the case on the building site. Furthermore, the acoustic properties of any component are subject to a certain scatter. The tolerance allowance is 5 dB for doors and 2 dB for all other components. The result is the so-called characteristic value $R_{w,R}$, which is used for sound insulation calculations.

Analysis for solid construction with flanking components of about 300 kg/m²
DIN 4109 supplement 1, 1989 edition, also contains resultant sound reduction indexes R'_w for components and assemblies in solid construction, where the mass per unit area of all flanking solid components is 300 kg/m² on average. If the flanking components deviate from this figure, allowances are added or subtracted. This is a purely pragmatic approach and is used in this form only in Germany.

Simplified analysis for frame construction
The requirement for the sound reduction index R'_w is deemed to be satisfied when the sound reduction index of the separating component $R_{w,R}$ and the sound insulation indexes of the flanking components $R_{L,w,R}$ are in each case at least 5 dB higher than the required value R'_w.

If, for example, a resultant sound reduction index of R'_w = 47 dB is achieved with a wall in dry construction, the product chosen must have a laboratory value (as specified by the manufacturer) of $R_{w,P}$ ≥ 54 dB (2 dB tolerance allowance plus 5 dB reduction for flanking transmissions).

Differentiated consideration of transmission paths for frame construction
This situation becomes somewhat more complicated when the sound insulation parameters of all the components involved in the sound propagation (R_w, $D_{n,f,w}$, etc.) have to be considered. The resultant weighted sound reduction index can then be calculated taking into account the geometric relationships and other constructional boundary conditions. DIN 4109 describes such calculation methods, e.g. for frame construction.

Application of the Euronorm DIN EN 12354
Even more detail has been added to this approach through the harmonised European standard DIN EN 12354, parts 1–6. In particular, the method of calculation for solid construction preferred in Germany up to now – based on measurements or characteristic values for taking into account flanking components with an average mass per unit area of 300 kg/m² and the use of correction factors for other average masses per unit area – will in future be replaced by a calculation model with 13 sound propagation paths. The calculations are based on parameters that describe the component, the weighted sound reduction index R_w without flanking transmissions. In addition, all flanking components and their interactions via their joints and junctions are considered in detail (p. 25, Fig. 1). This calculation model is described in DIN EN 12354 "Building acoustics", parts 1–6.

This series of standards was initiated by the European Construction Products Directive, which is intended to break down the trade barriers between the member states of the EU. Products for the construction sector are identified uniformly throughout Europe. In Germany the procedure for describing the technical

1

qualities of construction products is anchored in the Construction Products Lists of the DIBt (Deutsches Institut für Bautechnik – German Building Technology Institute). These are dynamic documents that list the methods of analysis for individual construction products based on European product standards or procedures for construction products not regulated by product standards in the form of attestation of conformity verifications with various levels of detail. Identification is by way of the CE marking (when a product standard is available) or national test certificates.

The specification of national requirements and measurements as input variables for the calculations are not affected by the European directive.
However, in light of deviations caused by different levels of workmanship, the method does not in practice always supply more accurate results than the method of calculation currently used according to DIN 4109. Furthermore, the DIN EN 12354 calculation methods have not yet been incorporated into German building legislation.

Calculating the sound insulation of composite components
Where the separating component consists of areas of different materials, e.g. partly glazed walls, facades with individual windows, the resultant sound insulation index $R_{w\,res}$ is calculated as follows (for two different surfaces, e.g. Fig. 2):

$$R_{w\,res} = -10\log\left[\frac{1}{S_1+S_2}\cdot(S_1\cdot10^{-0.1\cdot R_{w1}}+S_2\cdot10^{-0.1\cdot R_{w2}})\right]$$

where
S_1, S_2 = the surfaces
$R_{w,1}, R_{w,2}$ = the weighted sound reduction indexes of the components
If the sound insulation of component 2 is at least 15 dB below that of component 1

(e.g. doors in a wall), the equation can be simplified to the following:

$$R_{w,\,res} = R_{w2} +10 \cdot \log\left(1+\frac{S_1}{S_2}\right)$$

The resultant airborne sound reduction index is generally rounded off to a whole number.

Walls
Single-leaf solid walls
Single-leaf components are walls made from reinforced concrete or masonry and also lightweight internal walls made from plasterboard that consist of one leaf only. The rule here is the higher the mass per unit area of the wall, the higher is its sound insulation value.
From Fig. 3 we can see that weighted sound reduction indexes exceeding $R_w = 45$ dB can be achieved for mass per unit area values greater than about 200 kg/m². For heavyweight walls, the weighted sound reduction index without flanking transmissions increases by about 11 dB if the mass is doubled.
Greater deviations are possible when using densities not common in building (e.g. masonry unit densities > 2.2), in the case of leakage (e.g. at mortar joints in double-sided facing masonry that has not been very carefully erected) and when there are voids in the materials. Owing to additional resonance effects, this also applies to thermally insulating, lightweight, vertically perforated clay bricks. In such cases, the sound insulation and also the effects of flanking transmissions should be verified by the manufacturer by way of test certificates.

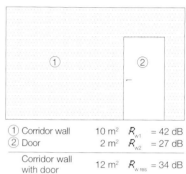

① Corridor wall	10 m²	R_{w1}	= 42 dB
② Door	2 m²	R_{w2}	= 27 dB
Corridor wall with door	12 m²	$R_{w\,res}$	= 34 dB

2

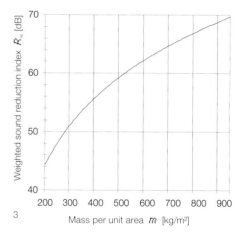

3

1 Construction of a masonry wall for testing purposes. After plastering and being allowed to dry out, the weighted sound reduction index is determined in a test according to ISO 140-3. The materials of the source and receiving rooms are kept separate by way of joints so that flanking transmissions are suppressed and only the sound insulation of the wall itself is determined.
2 Example of the calculation of the sound reduction index of combinations of components. The resultant sound insulation of combinations of components is always less than the maximum sound reduction index of the individual components.
3 Graph showing how the weighted sound reduction index R_w depends on the mass per unit area of a single-leaf component (without flanking transmissions). A 200 mm thick reinforced concrete wall with a density of 2300 kg/m³ (normal-weight concrete assumed for sound insulation calculations) has a mass per unit area of 460 kg/m² and achieves a sound reduction index of approx. $R_w = 58$ dB. If we double the wall thickness to 400 mm, the weighted sound reduction index climbs to $R_w = 69$ dB. (Note that this does not take into account the possible influence of flanking transmissions.)

The sound insulation of lightweight components is less than that of heavyweight components because their lower mass means they can be made to oscillate with a lower level of exciting energy. In addition, another physical effect occurs: a further decrease in the sound insulation above a certain frequency, the so-called coincidence frequency. This is the concurrence between the bending wave and the wavelength of the airborne sound, which means it is easier for the airborne sound field to excite oscillations in the component. Unfortunately, the coincidence frequency of lightweight components lies precisely in the frequency range relevant for building acoustics. The stiffer the material (low elasticity) and the thicker the wall, the lower is the coincidence frequency. The aim is either to keep this frequency as low as possible (heavyweight single-leaf walls ≥ 200 kg/m²) or above the critical frequency range (lightweight panels < 20 kg/m²). The worst case is when the coincidence frequency lies between about 200 and 2000 Hz because then the decrease in insulation at the coincidence becomes clearly perceptible. With such components, which include 80 mm thick gypsum building boards, for instance, not only is the sound insulation for the direct passage of sound low, but vibrations of heavyweight adjoining components are amplified, which leads to an increase in the radiation of sound.

Solid timber walls (glued laminated or cross-laminated timber) do not exhibit these disadvantageous acoustic phenomena owing to their layered construction and higher internal damping (Fig. 2). But owing to their low mass per unit area, the sound insulation for the direct passage of sound is limited. Where higher demands are placed on sound insulation (e.g. schools, preschool facilities), non-rigid wall linings are necessary.

1 Frequency-related sound reduction indexes of some solid components. The increase in the insulation value with the frequency is typical. In the case of the wall built from lightweight vertically perforated clay bricks, the drop in the sound insulation value between 1000 and 2000 Hz is clearly evident, which is attributable to the resonances of the clay bricks and in certain cases may become noticeable as a subjective drawback.

2 Frequency-related sound reduction indexes of lightweight wall constructions. In the case of plasterboard walls, the drop in the sound insulation value at the coincidence frequency is clearly evident. In this frequency range, the degree of insulation is already high and, in addition, the energy of typical noise sources is lower, which means that this is less significant subjectively. The poor sound insulation of the 75 mm thick plasterboard wall at low frequencies can also be seen (resonance of both leaves). As it is not a double-leaf construction and mass is lacking, the multi-ply timber element exhibits a much lower sound insulation value.

3 Double-stud walls, T-junction with isolating joint. Plasterboard-clad stud walls can achieve very high weighted sound insulation indexes of $R_w ≥ 60$ dB, depending on the boarding and the studs. Plasterboard walls as flanking components are not critical acoustically. If the boards are isolated on the room side, the flanking level differences exceed $D_{n,t,w} = 70$ dB.

4 If high demands are placed on sound insulation, e.g. between neighbouring offices, some form of bulkhead at least will be required in the void above a suspended ceiling. Example: bulkhead built from boards plus insulation on the main runner (beam) of a modular grid ceiling.

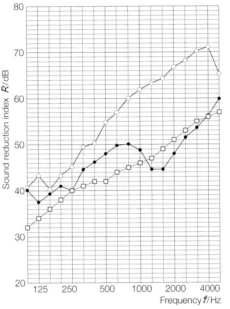

○——○ Plastered masonry wall made from calcium silicate bricks (240 mm, density class 2.0): $R_w = 58$ dB
●——● Plastered masonry wall made from lightweight vertically perforated clay bricks (300 mm, density class 0.9): $R_w = 48$ dB
□——□ Plastered masonry wall made from lightweight vertically perforated clay bricks (115 mm, density class 1.4): $R_w = 47$ dB
1

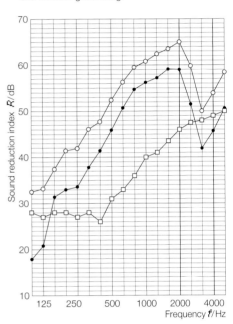

○——○ Lightweight plasterboard wall with two layers of plasterboard, 150 mm: $R_w = 53$ dB
●——● Lightweight plasterboard wall with one layer of plasterboard, 75 mm: $R_w = 44$ dB
□——□ Multi-ply timber element, 135 mm: $R_w = 37$ dB
2

Double-leaf solid walls

When building terraced and semi-detached houses, party walls that are structurally and acoustically separate are the norm. High sound insulation values of $R'_w = 62–72$ dB are possible when the cavity between the leaves is flawless (no acoustic bridges) and extends right down to the basement floor slab. To achieve this, the cavity must be at least 30 mm wide and filled with an insulating material of low stiffness. From the acoustics viewpoint, rigid polystyrene foam is unsuitable as cavity insulation. Good results can also be achieved by using heavyweight wall linings (see "Small rooms for music", p. 84).

Double-leaf lightweight walls

Lightweight separating walls in dry construction have a low mass; nevertheless, good to very good sound insulation values can be achieved with this form of construction. One reason for this is the double-leaf form with its intervening cavity, which together form a resonant system. Here, the leaves act as a mass and the cavity – owing to the compressibility of the enclosed air – as a spring. The following applies from about one octave above the resonant frequency:

$$f_R = 1700/\sqrt{d \cdot m'/2}$$

where
f_R = resonant frequency in Hz
m' = mass per unit area of one leaf in kg/m²
d = width of cavity in mm

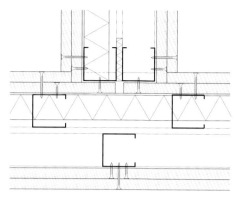

3

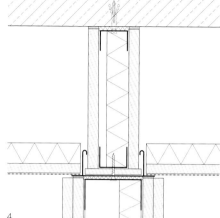

4

This means that higher sound reduction index values are achieved than is the case with a single-leaf component with the same mass per unit area. In the region of the resonant frequency, the sound insulation is poorer, and below the resonant frequency it is about the same. What this means for good sound insulation is that the resonant frequency should lie below the relevant noise spectrum. The resonant frequencies of typical plasterboard-clad stud walls lie between about 115 Hz (one layer of 12.5 mm plasterboard both sides, $m' = 8.5$ kg/m², cavity width 50 mm, total wall thickness 75 mm) and 58 Hz (two layers of plasterboard both sides, cavity width 100 mm, total wall thickness 150 mm).

It can be seen from the above resonant frequency equation that doubling the cavity width has the same effect on the resonant frequency (and hence the sound insulation) as doubling the mass per unit area. Filling the cavity completely with, for example, a mineral-fibre insulating material, which is necessary for thermal reasons anyway, ensures the necessary damping. Furthermore, the use of plasterboard, wood-based products or gypsum fibreboard in thicknesses of up to 15 mm is characterised by the fact that the coincidence frequency lies well above 1000 Hz. We speak of non-rigid leaves.

When using more than one layer of plasterboard or a similar material, the stiffness (and hence the coincidence frequency) does not alter because the layers are fixed only at individual points. Consequently higher sound insulation values can be achieved with two layers of 12.5 mm plasterboard on both sides than with one layer of 25 mm plasterboard each side. Another factor affecting the sound insulation that be can be achieved in the end is the connection between the leaves. Lightweight walls with timber studs behave less favourably than walls with 0.6 mm

thick CW or MW metal sections, which achieve an even better spring effect and hence decoupling of the leaves. Complete constructional separation and hence also the maximum sound reduction index is achieved by constructing a double-stud wall (applications: party walls, cinemas, music facilities).

There are also differences in the mass per unit area figures of the boarding. Whereas the weight of conventional plasterboard is about $m' = 8.5$ kg/m², the acoustic plasterboard varieties, which have been available for some years, weigh $m' = 10\text{--}12$ kg/m². Even heavier, and hence even better from the acoustics viewpoint, is gypsum fibreboard, where $m' = 12\text{--}15$ kg/m² for a 12.5 mm board. So depending on the application, different forms of construction can achieve weighted sound reduction indexes R_w between about 40 dB and about 72 dB. The dry construction industry provides plenty of information about forms of construction, details and sound reduction indexes. But we should not forget that the sound reduction indexes stated in many publications relate only to the direct passage of sound through the wall and the flanking transmissions still have to be considered as well.

The flanking walls are normally not critical when they are built as double-leaf lightweight walls in dry construction. Provided the boarding is interrupted completely at the junction with the separating wall, plasterboard walls exhibit very good flanking sound level differences of $D_{n,f,w} \geq 70$ dB (Fig. 3). So in terms of flanking transmissions, plasterboard walls are much better than lightweight solid internal walls.

Lightweight (non-rigid) wall linings
The good acoustic properties of dry forms of construction described above can also be employed as wall linings to improve the sound insulation of solid walls. The

clear width of the cavity between the boarding of a stud wall not connected to the solid wall should be at least 50 mm, depending on the application. Two layers of boarding should be used. The critical resonant frequency here is given by the following equation:

$$f_R = 1700/\sqrt{d \cdot m'}$$

where
f_R = resonant frequency in Hz
m' = mass per unit area of wall lining
  in kg/m²
d = width of cavity in mm

Demountable partitions
The demountable partitions that are being used more and more, especially in office buildings owing to the flexibility they provide, behave similarly to plasterboard-clad stud walls. In acoustic terms, however, there are differences in the quality because the different manufacturers of these prefabricated systems employ different forms of construction. Where sound insulation requirements are important, the tender documents should therefore always specify the weighted sound reduction index, which should be verified by the supplier by way of a test certificate. Feasible values are in the range $R_{w,P} = 40\text{--}50$ dB. Demountable partitions often include large areas of glazing. Here again, there are differences in terms of quality and cost. For example, double glazing with a cavity width of, for example, 60 mm and an intervening reveal made from an absorbent material can certainly achieve weighted sound reduction indexes of $R_{w,P} = 45$ dB. Single glazing cannot achieve values higher than about $R_{w,P} = 37$ dB.

1

1 Doors with poor sound insulation often lead to subsequent measurements in the as-built condition. Using a measuring loudspeaker in the shape of a dodecahedron, a sound source is created at the door which is measured directly in front of the door and also directly behind it.
2 Floating screed with strips of insulation around the edges. In order to avoid structure-borne sound bridges, such strips of insulation should not be cut back until after laying the floor finishes.

Doors and movable room dividers

Doors

Doors, too, are in many situations expected to provide adequate sound insulation. But there is no other building component where the discrepancy between intended and actual sound insulation can be so significant. The reasons for this are, for example, distorted leaves, and hence the lack of a consistent contact pressure on the seals between door and frame, the poor quality of the seal at floor level (if provided at all) and also the lack of packing behind the frame members. The wording of tender documents should be unambiguous. If, for example, a weighted sound reduction index of $R_w = 32$ dB is required, the specification should read as follows:

- Weighted sound reduction index of doors in as-built, functioning condition to DIN 4109, measured in tests to ISO 140-3: $R_{w,P} \geq 37$ dB

The suffix "P" makes it absolutely clear that that this is a value measured in a laboratory test, which is 5 dB (tolerance allowance to DIN 4109) higher than the value required in situ.
If the expression "as-built, functioning condition" is omitted, it can happen that the weighted sound reduction index refers to the door leaf only; the complete system with seals and frame does not then comply with the values required. When the weighted sound reduction index required is $R_w \geq 32$ dB, care must be taken in the design and on the building site to ensure that this value is actually achieved. Dimensionally stable frames and sound-insulating leaves are just as important here as the packing behind and sealing to the frames. If the seals do not close tight on all sides, even a high-quality door leaf cannot achieve a good level of sound insulation.

All-glass doors with slender frames and automatic door bottom seals cannot achieve values higher than $R_{w,P} \approx 37$ dB. T30 doors can provide sound insulation values of up to $R_{w,P} \approx 50$ dB, T90 doors up to $R_{w,P} \approx 42$ dB.
A double door generally exhibits lower insulation values than a single door with the same construction.
Where it is important to achieve especially high sound insulation values, pairs of doors can be used, but emphasis must be placed on good workmanship and the layout must permit this type of solution. This arrangement involves using two doors, one on each side of the wall. Where the distance between the doors exceeds about 1.5 m, we speak of an airlock. The ceiling, possibly also the walls, between the doors should be lined with a sound-absorbent material. With this solution it is possible to achieve, for example, using one $R_{w,P} = 32$ dB door and one $R_{w,P} = 37$ dB door, a weighted sound reduction index of $R_w \approx 50$ dB in total in the as-built condition.
When self-closing, heavyweight doors, e.g. steel fire doors, are necessary, the noise of the door closing can be avoided by using an automatic door closer with a hydraulic or air check, or door fittings with a damping effect. Maximum sound pressure levels for the closing action of self-closing doors in the nearest rooms requir-

ing sound insulation and not in the same unit of occupation are laid down in building legislation, e.g. for housing $L_{AF,max} \leq 30$ dB(A).

Industrial doors

Industrial doors are treated in a similar way to normal doors. The sound reduction indexes to be expected in the as-built condition range from about 15 dB for roller-shutter or overhead doors without special acoustic measures to $R_w \approx 45$–50 dB for heavyweight sound-insulating doors.

Movable room dividers

Movable room dividers are available with weighted sound reduction indexes of up to about $R_{w,P} = 55$ dB. Here again, a tolerance allowance of 5 dB should be applied to obtain the value possible in the as-built condition. Only by taking into account the flanking components and designing and building with care is it possible to achieve a sound insulation value of $R'_w = 45$ dB. In the case of flanking components that are disadvantageous in terms of acoustics, e.g. a continuous floating screed, the resultant sound insulation is far lower. In the end it is necessary to specify plausible sound insulation requirements to match the use of the rooms to be divided and ensure through detailed planning of all possible transmission paths – which means

T2: Construction advice for doors, depending on the weighted sound reduction index R_w required in the bldg.

R_w[1] [dB]	$R_{w,P}$[2] [dB]	Door leaf	Seals	Remarks
27	32	Any form of construction $m' \approx 25$ kg/m² approx. 40 mm thick	Frame seals, automatic door bottom seal	Fairly easy to realise, automatic door bottom seal can run over carpet if necessary.
32	37	Multi-layer construction[3] $m' \approx 35$ kg/m² approx. 40–55 mm thick	Frame seals, automatic door bottom seal	Use of tested door systems recommended, careful design and construction necessary.
37	42	Multi-layer construction[3] $m' \geq 40$ kg/m² approx. 60–70 mm thick	Double rebate seals, automatic door bottom seal combined with raised threshold	Use of tested door systems essential, careful design and construction necessary.

[1] Doors in as-built, functioning condition
[2] Doors in functioning condition in test laboratory
[3] or double-skin, special sound-insulating door leaves

2

T3: Impact and airborne sound insulation of reinforced concrete suspended floors[1] based on DIN 4109 supplement 1

	Mass per unit area m' [kg/m²]	$L'_{n,w,R}$ [dB]	$R'_{w,R}$ [dB]
120 mm	280	81 - ΔL_w	48 (+6)[2]
160 mm	370	76 - ΔL_w	51 (+5)
200 mm	460	73 - ΔL_w	54 (+4)
250 mm	575	70 - ΔL_w	57 (+3)

[1] The values are valid for solid construction and an average mass per unit area of the flanking components $m' = 300$ kg/m². If the flanking components are heavier or non-rigid (plasterboard walls), the values tend to be more favourable from the acoustics viewpoint.
[2] Values in brackets: improvement due to floating floor finishes with $\Delta L_w \geq 24$ dB.

all trades must be involved – that these requirements can be attained in the as-built condition.

Suspended floors and floor finishes

Solid suspended floors
The airborne sound insulation of solid suspended floors, in particular those of reinforced concrete, is similar to that of solid walls. In the case of a hollow clay block or similar floor, the insulation can be marginally inferior to that of a solid slab with the same mass per unit area. Besides the airborne sound insulation, it is the impact sound insulation that plays an important role in the case of floors. Generally, the impact sound insulation of structural floors without any finishes is low, even with a high mass (high $L'_{n,w}$ value), as can be seen in Tab. T3.

Floating screeds
The impact sound insulation can be considerably improved by laying a screed on top of a suitable impact sound-reducing layer (floating screed).
Together with the structural floor, a floating screed forms a double-leaf resonant system (similar to a wall plus wall lining). Here, too, the floating screed develops its improving effect above the resonant frequency, which depends on the mass per unit area of the screed and the stiffness of the layer of insulation. For impact sound insulation, the dynamic stiffness should lie between 10 and about 50 MN/m³.

Moreover, adding a floating screed to a solid suspended floor also enhances the airborne sound insulation. In contrast to reducing the impact sound, however, the improvement depends on the mass of the structural floor and lies between about 3 dB (heavyweight 250 mm deep reinforced concrete slab) and 6 dB (e.g. lightweight reinforced concrete ribbed slab with 100 mm slab).
Mineral-fibre or elasticised polystyrene boards are the normal choice for impact sound insulation. Natural products, e.g. wood-based insulating or coconut-fibre boards plus various recycled materials are also in use. In the CE marking scheme for impact sound insulation products, SD 20 means a dynamic stiffness of max. 20 MN/m³. The impact sound reduction in conjunction with a solid floor is in the range $\Delta L = 20$–30 dB (Tab. T4). Besides the acoustic properties, the maximum imposed loads are also important for the design. For floating screeds and imposed loads of up to 2 kN/m² (housing), also up to 5 kN/m² (places of assembly), a dynamic stiffness of max. 20 MN/m³ (SD 20) is a good acoustic value for 20–30 mm thick insulating boards with low compressibility.
Thinner layers of impact sound insulation are generally stiffer than thicker ones. There is a risk of structure-borne sound bridges (i.e. rigid connections between screed and structural floor) with layers of insulation < 10 mm thick because in such cases general building debris or unevenness of the structural floor might tend to punch through the insulation.
Such acoustic bridges reduce the effectiveness of the impact sound insulation substantially, lead to more re-radiation of airborne sound and very often turn out to be the cause of damage and complaints. For these reasons, a number of details are investigated in the following.
It should be obvious that continuous strips of insulating material, e.g. PE foam or mineral-fibre products, are required not only alongside walls, but also around pipe penetrations, door frames, radiator brackets, etc. When laying self-levelling screeds, strips of foil-backed insulating material around the edges ensure that the work is carried out properly because these are bonded to the separating layer to form a tank-like construction which prevents the screed from flowing underneath.

The strips of insulating material should project at least 30 mm above the screed and should not be cut back until after the floor finishes have been laid in order to prevent structure-borne sound bridges caused by debris in the joints or when laying the floor finishes (e.g. tile adhesive/grout, wood shavings, carpet adhesive, etc.). It is advisable to include the cutting-off of excess insulating material in the floor finishes contract.
At the doors to rooms with acoustic requirements, the screed must be interrupted by building in a strip of insulating material. Cutting a slit with a trowel is not an adequate way of preventing a structure-borne sound bridge.
Where lightweight walls are built off continuous floating screeds, the maximum weighted sound reduction index feasible is $R'_w \approx 35$–40 dB because the sound propagates horizontally within the screed (similar to flanking thin solid walls with a low mass).
Another effect should be pointed out here: the resonant frequency of floating screeds usually lies in the range 60–80 Hz – the range in which vibrations due to railborne

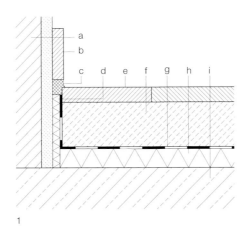

1 Detail of floating screed at junction with wall
  a Separating wall
  b Skirting board
  c Permanently resilient sealant
  d Strip of insulating material
  e Floor covering
  f Cement screed
  g Separating layer, e.g. PE sheeting
  h Impact sound insulation, e.g. mineral-fibre insu-
    lating material to EN 13162, application type
    DES-sg, compressibility CP2, dynamic stiffness
    SD ≤ 20 MN/m³
  i Structural floor
2 Despite their low mass, cross-laminated timber or
  similar timber floors can be used to achieve the
  sound insulation requirements for housing or
  schools, for instance, provided suitable floating
  screeds are laid on the floors.

traffic also occurs. Where a new construction project is situated in the vicinity of a railway line (100 m or nearer, possibly even further away where the subsoil is rock), the possible structure-borne sound transmissions should be investigated within the scope of a vibratory effects study. Otherwise, it can happen that the increased oscillations in the screed in the region of the resonance lead to disturbing low-frequency sound emissions and vibrations.

Dry subfloors
Dry subfloors are especially popular in refurbishment projects because very often low weight and a low depth of finishes is essential. Because of the low mass of products made from, for example, 23 mm gypsum fibreboard, E100 particleboard, etc., plus the higher stiffness of the impact sound insulation needed for a dry subfloor, the impact sound reductions are less than those achieved with floating wet screeds and lie in the range of 10–18 dB. In addition, dry subfloors are less effective with low frequencies.

Resilient floor coverings
Carpeting, linoleum on a "Korkment" underlay and real wood laminate flooring on a thin layer of impact sound insulation also exhibit an insulating effect. The values given in Tab. T4 are valid when laying such floor coverings directly on the structural floor. If they are laid on a properly installed floating screed, the additional improvement due to the floor covering is, however, only 0–2 dB.
According to DIN 4109, 1989 edition, resilient floor coverings may not be taken into account when verifying the impact sound insulation in housing because they can be changed, even removed completely, by the occupants.

Noise of walking
The impact sound insulation does not describe what sounds ensue in the room itself when walking across the floor covering with hard-soled footwear. This can be very disturbing, especially when thin, and hence lighter, laminate finishes (with low internal damping) are simply laid on the subfloor. In the meantime, laminate floor coverings are available in which this effect is reduced. But as different groups within the industry have introduced different acoustic parameters, an objective comparison is difficult.

Hollow floors
Hollow floors and raised access floors enable the flexible routing of pipes and cables in the floor. Their acoustic properties essentially depend on their construction, but also on the floor covering. One particular problem can be the transmission of impact sound in the horizontal direction when using hard floor coverings because the walls are normally erected on top of the hollow floor, leaving an uninterrupted void underneath. In order to reduce the horizontal sound propagation, various interruptions to the hollow floor construction below partitions have been tried out, especially for hard floor coverings. However, in practice the effect has frequently been found to be inadequate because the interruption is incomplete.

In order to reduce the impact sound propagation in the vertical direction, an underlay can be placed beneath the framing to the hollow floor to dampen the structure-borne sound transmissions. As the testing methods have been substantially altered since 2000, manufacturers' data or test certificates issued before that should no longer be used when making an assessment. The old testing methods typically led to higher values. The new testing methods are carried out according to DIN EN ISO 140-12 or, since 2006, according to DIN EN ISO 10848-2.

Timber joist floors
Timber joist floors are often found in older buildings, but this form of floor construction is also enjoying a comeback in the building of new detached homes.

The difference between timber joist floors and solid floors is their much lower mass. It is therefore easier to excite oscillations in them. And the noise control of older suspended floors no longer corresponds to modern expectations of sound insulation. Typical values without a resilient floor covering and with a rigidly attached ceiling are R'_w = 44–48 dB or $L'_{n,w}$ = 66–70 dB. Furthermore, the improvement in impact

T4: Reduction in impact sound insulation ΔLw for floor finishes on solid structural floors

Overall construction depth [mm]	Screed or floor covering	Mass per unit area of screed m' [kg/m²]	Stiffness of impact sound insulation s' [MN/m³]	ΔL_w [dB]
70–100	Cement or calcium sulphate screed[1]	≥ 70	20	28
60–90	Cement or calcium sulphate screed[1]	≥ 70	40	24
50–70	Asphalt[1]	≥ 45	30	24
30–50	Dry subfloor[1]	≥ 25	40	21
25	Real wood laminate flooring	–	50	18
4–8	Carpeting	–	–	18–26
3–5	Linoleum	–	–	3–15

[1] with hard floor covering. The impact sound insulation increases by max. 2 dB when using a resilient floor covering..

2

sound insulation gained with conventional wet screeds is much poorer than when they are used in conjunction with solid suspended floors, especially for low frequencies. Pugging usually contributes only very little to improving the sound insulation of older suspended floors when the ceiling is rigidly fixed to the joists because most of the sound is transmitted via the joists.

Which measures are feasible when refurbishing an existing building depend on many factors. If the ceiling can be removed – but sometimes preservation orders do not allow this –, a new ceiling consisting of two layers of 12.5 mm plasterboard can be suspended below the timber joists on resilient brackets or channels. The ensuing void is then packed with min. 80 mm thick mineral wool. Where it is possible to increase the depth of the floor and to add ballast, one option is to lay concrete bricks (= non-rigid ballast) below a floating screed or dry subfloor. Alternatively, a heavyweight wet screed can be installed. There is also the possibility of laying strips of resilient insulating material on the levelled joists and laying a screed over these on boarding (= permanent formwork). Such measures generally achieve the minimum sound insulation requirements customary for housing. If the different upgrading measures are combined, much higher sound insulation figures can be achieved, provided the flanking transmissions are also checked at the same time ("Case studies", pp. 88–89).

Stairs
A weighted normalised impact sound level of $L'_{n,w,R}$ = 65 dB can be expected with solid stair flights connected rigidly to the enclosing walls (where $m' \geq$ 380 kg/m²). Much better values are possible when the structure-borne sound bridges are elimi-

nated, i.e. the stairs are separated from the walls via joints and the stair flights are supported on resilient bearings at the floors and landings.

Lightweight steel stair constructions can cause disturbing structure-borne sound transmissions, especially when these adjoin the floors and/or walls to rooms requiring sound insulation. One remedy is to use fixings that dampen the structure-borne sound transmissions.

Roof structures
The airborne sound insulation of roofs is usually only relevant when the level of external noise is high, or in the case of venues.
The weighted sound reduction index of a pitched roof with insulation over or between the rafters is $R_w \approx$ 42–52 dB, depending on the particular construction. Lightweight roofs made from steel trapezoidal profile sheeting can achieve similar values; indeed, even higher values are feasible. Reinforced concrete roofs essentially behave like reinforced concrete floors, i.e. the mass per unit area is important.
On the other hand, care is needed with insulation over the rafters which is made from non-porous, stiff materials (e.g. PUR) because they can lead to an increase in the lanking transmissions via the roof, e.g. in multi-occupancy residential buildings. With metal roof coverings, there may be a risk that the noise of rainfall could create a disturbance and in the case of tent-like textile roofs the very low weighted sound reduction index of, typically, $R_w \approx$ 10–15 dB must be considered as well as the noise of rainfall, which can lie in the region of 60–70 dB(A). Such roofs are therefore less suitable for venues, classrooms or workrooms. When roofing over circulation zones, the potential for disturbance should be weighed against the usage profile.

Windows
The sound insulation characteristics of a window depend essentially on the quality of its glazing. The influence of the frame is generally minimal up to a weighted sound reduction index of $R_w \approx$ 45 dB. This also applies to post-and-rail facades for the direct passage of sound from outside to inside (but not with respect to the flanking transmissions). In the case of newer windows with opening lights, the functional joints are less of a problem because they have to exhibit a high airtightness for thermal reasons anyway.
Windows can be assigned to sound insulation classes (SSK) to VDI 2719, as shown in Tab. T5 (p. 36).
Single glazing can achieve weighted sound reduction indexes of R_w = 30–35 dB, depending on thickness, and 10 mm thick laminated glass a value of R_w = 37 dB. An approx. 0.4 mm thick sound-deadening sheet based on polyvinyl butyral (PVB) can be used as the interlayer in laminated glass, which in the meantime is being preferred to laminated glass using casting resin.

Weighted sound reduction indexes of up to R_w = 50 dB are possible with insulating glass units. For sound insulation class 3 and higher, the two panes should be of different thicknesses in order to avoid identical coincidence frequencies.
Coupled windows consisting of two windows spaced approx. 100–300 mm apart can achieve values of $R_{w,P}$ = 52–62 dB. Such windows are very common in music studios, for example (window between recording and listening rooms).

For glazing in particular, so-called spectrum adaptation terms C, C_{Tr} are important in addition to the weighted sound reduction index R_w, which then reads as follows, e.g. $R_w (C; C_{Tr})$ = 41(0; -5) dB. These spectrum adaptation terms to ISO 717-1 take into account the spectral composition of different types of sounds (e.g. index Tr = traffic). The insulation of a window against road traffic noise increases with the sum of the weighted sound reduction index plus the spectrum adaptation term $R_w + C_{Tr}$. When building alongside busy roads, this aspect should be taken into account during the design work and tendering procedure.

Insulating glass units exhibit resonant frequencies of about 120–250 Hz, depending on the width of the cavity and the thickness of the panes. As critical components of road traffic noise are to be found in this range, the spectrum adaptation terms turn out to be negative and normally lie in the range C_{Tr} = -4 to -9 dB. Compared to double glazing with the same thickness and mass, triple glazing has poorer acoustics (approx. 2–3 dB) because the additional pane generates an unfavourable intermediate resonance. Filling the cavity with a gas influences not only the thermal insulation, but the sound insulation as well. Instead of an air or pure argon filling, sulphur hexafluoride SF6 was added until very recently, which brought about an improvement of R_w = 2–3 dB. But for environmental reasons, SF6 is no longer used.

Thermal insulation materials
The thermal insulation materials built into external walls or roofs can either improve or worsen the sound insulation, depending on the resonant frequency of the system. For instance, thermal insulation composite systems with stiff insulating materials (e.g. so-called mineral lamella boards or non-elasticised rigid polystyrene foam) in conjunction with 5–20 mm thick coats of render exhibit resonant frequencies of 200–500 Hz. The sound insulation in this frequency range is then lowered, which has an effect on the weighted sound reduction index. Where higher external noise levels and/or high requirements

regarding a low internal level are relevant, this aspect should not be neglected.
In older buildings, the internal thermal insulation in recesses for radiators, consisting of plastered lightweight wood-wool boards, represents another disadvantage. Owing to the stiffness of the layer of insulation, the resonance lies in an unfavourable frequency range.
Facades with an intermediate ventilation cavity do not generally present any acoustic problems and can even achieve a marked improvement in the sound insulation.

Building services
The building services relevant to acoustics include:
- lifts
- heating systems, heat pumps
- ventilation systems
- automatic door mechanisms
- drinking-water and waste-water pipework
- wash-basins, WCs, urinals and showers

Requirements
DIN 4109 specifies the requirements for the maximum permissible sound pressure levels due to building services (Tab. T6). VDI 2081 supplement 1 specifies recommended values for ventilation systems (Tab. T7). Besides the A-weighted level, frequency-related figures are specified for particularly demanding acoustic situations.
Building services are assessed according to the maximum sound pressure level occurring $L_{AF,max}$. There is a separate measuring specification for water pipework which specifies how the installation noise level L_{In} in dB(A) is calculated from the individual noise components.

T5: Sound insulation classes to VDI 2719 and weighted sound reduction indexes (minimum values in as-built condition). Typical constructions for glazing and windows with insulating glass

Class	R_w [dB]	Glazing [mm]	Window with insulating glass [mm]	
1	25	TSG 4	any	
2	30	TSG 8, LSG 6	TSG 4 / cavity 8 / TSG 4	d: 16
3	35	LSG 8	TSG 8 / cavity 16 / TSG 4	d: 28
4	40	LSG-PVB 16	LSG-PVB 8 / cavity 12 / TSG 10	d: 30
5	45	–	LSG-PVB 12 / cavity 24 / LSG-PVB 8	d: 44
6	50	–	Special construction, e.g. coupled window	d: 100

TSG: toughened safety glass; LSG: laminated safety glass; LSG-PVB: with acoustic PVB interlayer; cavity: space between panes filled with argon or air; d: total thickness

T6: Maximum permissible sound pressure levels in rooms requiring sound insulation, caused by building services, to DIN 4109, amendment A1, January 2001

Source of noise	Characteristic sound pressure level	Living rooms and bedrooms	Classrooms and workrooms
Water pipework[1]	Installation level L_{In}	30	35
Other building services[2]	Max. sound pressure level $L_{AF,max}$	30	35

[1] Individual brief peaks, which occur when operating fittings, are not taken into account.
[2] Values 5 dB higher are permissible for ventilation systems when constant noises without conspicuous individual pitches are concerned.

Ventilation systems

The sound insulation for ventilation systems generally comprises measures for the airborne and structure-borne sound insulation of the plant plus measures for reducing the sound propagation via the ducts. Information about the machinery is supplied by the frequency-dependent sound power level (broken down according to equipment emissions, duct supply and extraction sides).

If the ventilation ducts do not branch off directly from the corridors into the rooms, but instead run from room to room, the risk of "crosstalk" is increased. In addition, the wall penetrations represent acoustic weak spots.

Some refrigeration units generate very high sound pressure levels of 85–100 dB(A), with a high proportion of structure-borne sound. Accordingly, wherever possible, plant rooms containing such machinery should not be placed in the vicinity of rooms requiring sound insulation – otherwise elaborate, expensive sound insulation measures for walls and floors will be necessary. Elastic bearings are available for such plant. Depending on the degree of decoupling required, elastic mountings, possibly with an intermediate foundation, right up to special acoustically designed multiple elastic bearings may be needed. In such situations, it should be remembered that these bearings require at least 100 mm of additional height, sometimes up to 400 mm.

Lifts

The operation of lifts and the movement of the lift cars cause scraping and/or impact noises in the lift shafts, which can be transmitted to other building components. If the lift shaft adjoins rooms requiring sound insulation, reinforced concrete walls at least 250 mm thick will be required. As a rule, lift motors are mounted on elastic bearings. Design advice can be found in

VDI 2566 "Acoustical design for lifts with/ without a machine room".

Water pipework

According to DIN 4109, walls to which water pipes are fixed should either make use of non-rigid dry construction or wall linings, or, where solid walls are concerned, should have a mass per unit area of at least 220 kg/m^2, e.g. a 115 mm thick masonry wall plastered both sides and built from units of density class 1.8, or a 175 mm thick masonry wall plastered both sides and built from units of density class 1.2 or higher.

To limit the noise from pipework fittings, use only fittings of DIN 4109 fittings group I. Such fittings have all been tested for their noise emissions. Allocation to fittings group I is indicated by a test mark, e.g. P IX 123/I.

Limiting the water pressure by using a pressure reducer also helps to reduce the noise from pipework fittings.

Sounds from waste-water pipes are normally due to falling and impact noises around bends and elbows. If waste-water pipes pass through rooms requiring sound insulation, the use of heavyweight materials such as SML cast pipes (socketless cast iron waste-water system), heavy plastic pipes or a casing of sound-insulating material brings benefits. Alternatively, or in addition, it is possible to enclose the pipes in a double layer of 12.5 mm plasterboard.

T7: Recommended values for the sound pressure levels from ventilation systems to VDI 2081 (extract)

Function of room	L_{eq} [dB] high – low requirement
Concert hall	25–30
Rest room Break room Reading room Hotel bedroom	30–35
Conference room Classroom Seminar room Lecture theatre Museum	35–40
Open-plan office Booking hall, banking hall	45–50
Sanitary facilities	45–55

Noise control in urban planning

In many construction projects it is not only the noise control measures within the building that need to be considered, but also protection against external noise. This involves checking which sound immissions from outside impact on the building envelope, e.g. when building new housing adjacent to a busy road. However, the issue of noise, and protection against excessive noise levels, is not only important in individual construction projects, but also in the overriding urban planning.

After the fumes from vehicle exhausts, noise is the most significant environmental pollutant in the urban environment, with road traffic noise responsible for most complaints concerning noise.

But an increasing source of noise pollution is that generated in conjunction with leisure activities, such as sports events in football and athletics stadiums, the use of swimming pools for recreation, or the arrival and departure of cars at restaurants, public festivals and other events.

In established urban structures, conflict situations can be caused by an increasing amount of traffic, commercial uses or even loud individual events. Conflicts also arise when planning new factories or leisure facilities, also amenities that will generate traffic in the vicinity of housing requiring protection from noise, or vice versa. The same applies when both buildings requiring protection from noise and also potential noise producers are being planned.

In researching the effects of noise, we distinguish between its effects on our health as well as our physical and social well-being.

According to data provided by Germany's Federal Environment Agency, residents living alongside roads where the average sound level during the day exceeds 65 dB(A) have a 20 % higher risk of suffering a heart attack. An average daytime level of 55 dB(A) measured on road and rail routes is enough to impair communication and well-being with the windows even partly open. Sleep disrupted by noise frequently leads to serious health disorders. An average level of 45 dB(A) at night due to road traffic is particularly disruptive.

Continuous noise is a burden on our general physical well-being and our immune system. Depending on an individual's particular constitution, physical reactions can be triggered even at very low levels and lead to headaches, drowsiness and excessive irritation.

Noise curtails the quality of life for many people considerably. People feel particularly disturbed during their periods of recreation and relaxation, i.e. during their leisure time at weekends and after work.

Noise can also cause changes in social structures. According to studies by the Federal Environment Agency, urban noise contributes noticeably to the tendency for households with a higher proportion of disposable income to relocate to the suburbs, whereas the less prosperous become concentrated in areas with a high level of noise pollution.

Based on such findings, maximum permissible noise levels are specified for urban planning. Some of these are internal levels, but others relate to the outside world so that a certain quality of life is guaranteed for gardens, balconies and also public spaces.

Assessment of sound immissions – legal basis

Noise immissions are assessed according to legal requirements by using limit, recommended or reference values. Laws, statutory instruments and directives are frequently based on different source types. What this means is that road traffic, rail traffic, air traffic, industry and commerce, also sport and leisure activities all have their own, different regulations and assessment methods. Two important overriding legal documents are the German Immissions Control Act (*Bundes-Immissionsschutzgesetz*) and the EU's Environmental Noise Directive.

European Environmental Noise Directive
The aim of the European Union's Environmental Noise Directive (END) is to reduce the exposure of persons to environmental (i.e. ambient) noise (= noise abatement). It is for this reason that environmental noise has to be assessed uniformly for primary transport routes and urban agglomerations with more than 250 000 inhabitants. The harmful effects should be prevented wherever possible or at least reduced, or precautionary measures taken to prevent noise in the first place. In doing so, the public will be informed and made aware of the issues.
The environmental noise exposure is to be recorded in the form of strategic noise maps (sound immissions plans) which will create the basis for overriding noise abatement planning. Furthermore, tables of data concerning, for example, violation of relevant limit and guidance values will be drawn up.
Based on this, local authorities will be obliged to prepare action plans in which the noise problems and the effects of noise are dealt with and noise abatement measures are shown. Noise maps and action plans are to be reviewed and updated if necessary every five years.

1　Apartment blocks in Munich, 2000,
　Fink + Jocher
　Glazed loggias provide additional protection
　against noise.

German Immissions Control Act

In Germany the most important law for protecting people against harmful environmental effects – and hence also noise – is the *Bundes-Immissionsschutzgesetz* (BImSchG). Together with associated regulations and administrative documents, this act regulates the following aspects in particular:

- (cl. 4ff.) Protection against noise and the operation of installations with specifications regarding recommended immissions values in the German Technical Rules for Protection Against Noise (*TA Lärm*).
- (cl. 38) Limit noise values for vehicles in conjunction with the German Road Traffic Act.
- (cl. 41 – 43) Protection against noise in connection with the building of new or major alterations to existing transport routes; the specification of limit values and methods of calculation are specified in the German Traffic Noise Control Act (*Verkehrslärmschutzverordnung*) (16th BImSchV).
- (cl. 47) Local authorities' obligation to measure, assess and if necessary reduce noise exposure levels for areas in which harmful noise effects are present or are to be expected in the form of noise abatement plans.
- (cl. 50) The arrangement of land-use areas in such a way during planning that harmful environmental effects are avoided.

Regulations for urban planning

Noise control must be taken into account to a reasonable extent in urban planning measures (land-use plans, the legal basis for which is embodied in the Federal Building Code) in an attempt to prevent noise problems from the outset.

DIN 18005 "Noise abatement in town planning" represents a crucial basis for assessment in this context. This standard contains, in supplement 1, acoustic guidance values depending on the use of the land (Tab. T1). Keeping to or below these values is desirable in order to satisfy the expectations regarding reasonable protection against noise appropriate to the character of the land use.

However, this standard is not a statutory instrument. The guidance values given in the standard should therefore be considered carefully.

For example, in areas already burdened with noise pollution, especially in the case of existing buildings and established urban structures, existing transport routes and conflict situations, it is sometimes impossible to keep to the guidance values. Where deviations from the guidance values are made because other matters are more important and this is considered reasonable, the deviation should be compensated for, if possible, by measures at the building itself (e.g. suitable building orientation and plan layout, sound insula-

tion measures, especially for bedrooms) and the design checked against the legal requirements.

Regulations for road and rail traffic

The provisions of the Traffic Noise Control Act (16th BImSchV) apply when implementing the Immissions Control Act in conjunction with the building of new or major alterations to existing public road or rail (including tram) routes.

When building or carrying out major changes, it should be ensured that the assessment level does not exceed the corresponding immissions limit value (Tab. T2).

The nature of the building works and the areas are determined by the stipulations of the development plans. If no development plan is available, the building works are to be assessed according to their need for protection against noise.

If the immissions limit values are exceeded, the Traffic Noise Abatement Measures Act (24th BImSchV) is one of the statutory instruments that regulates the nature and scope of the noise abatement measures necessary at the building (passive noise control measures). However, the effects of existing transport routes are not taken into account. The upshot of this could be that the facades and sound-insulating windows designed according to the 24th BImSchV are inadequate according to research into the effects of noise.

There is no legally stipulated right to expect an upgrade to the sound insulation of a building alongside a loud road or railway. Instead, there exists the noise abatement programme for publicly owned trunk roads and railway lines. Noise control upgrading measures are carried out only when sufficient funds are available (Tab. T2).

T1: Acoustic guidance values for urban planning [dB(A)] to DIN 18005

	Day	Night[1]
Purely residential, holiday homes/chalets	50	40/35
General residential, small estates, camping sites	55	45/40
Special residential	60	45/40
Villages and mixed uses	60	50/45
Inner-city and commercial	65	55/50
Special areas, providing protection is needed, depending on type of use	45–65	35–65
Cemeteries, allotments, parks	55	55

[1] Lower values apply for noise due to industrial, commercial and leisure activities.

German Air Traffic Noise Act
This act, revised in 2007 and valid throughout Germany, exists to protect people living in the vicinity of airports and aerodromes. It contains limit values for noise control zones associated with the building of new and expansion of existing facilities plus the requirements regarding passive noise control for developments requiring protection.

Industrial and commercial operations
The German Technical Rules for Protection Against Noise (*TA Lärm*) should be used to assess industrial and commercial operations according to the German Immissions Control Act (BImSchG).
However, the following do not fall within the remit of these rules:
- Shooting ranges where weapons exceeding 20 mm calibre are used (immissions due to shooting are to be determined according to directive VDI 3745 "Assessment of shooting noise")
- Open-cast mining and seaport transshipment terminals
- Agricultural facilities not subject to approval permits
- Facilities for social purposes
- Sports facilities, other leisure facilities
- Building sites

The rules contain recommended immissions values (Tab. T3) that depend on the classification of the area. These values relate to the sum of all noise immissions from industrial sound sources acting at the immissions point. Noise immissions due to other types of sound sources (e.g. traffic noise, sports and leisure activities) must be considered separately.

Sports and leisure facilities
It is necessary to refer to the Sports Facilities Noise Abatement Act (18th BImSchV) when establishing and operating sports facilities. The assessment of leisure facili-

T2: Limit and recommended values for noise control alongside traffic routes [dB(A)]

	Precautionary[1] Day	Precautionary[1] Night	Refurbishment[2] Day	Refurbishment[2] Night
Hospitals & similar	57	47	70	60
Residential	59	49	70	60
Mixed uses	64	54	72	62
Commerce/industry	69	59	75	65

[1] According to 16th BImSchV for new and major alterations to roads or railway lines.
[2] For roads and railway lines for which the federal government is responsible, provided funds are available.

ties is handled differently in the different German Federal states.
Sports facilities also includes amenities nearby and closely associated with the operation of the sports facilities (e.g. clubhouse, car park).

Assessment level
The aforementioned limit, recommended and reference values always refer to the so-called assessment level L_r.
Here, the long-term average sound level determined over a defined assessment time L_{eq} (average level, p. 10) is adjusted by factors laid down in standards. The assessment level L_r is therefore made up of the long-term average sound level L_{eq} for the respective assessment time and additions/reductions K_i:

$$L_r = L_{eq} + \Sigma K_i$$

For example, road traffic noise attracts an "intersection surcharge" to allow for the increased disturbance in the vicinity of traffic lights. Surcharges are also applied when noises contain many impulses, pitches or information because these increase the disturbing effects. As rail traffic noise is generally regarded as less disturbing, a "rail bonus" of 5 dB applies in Germany. This means that the assessment level lies 5 dB below the long-term average sound level L_{eq}.

Assessment time
The assessment time "day" is generally taken to be 16 hours, from 6 a.m. to 10 p.m., and the assessment time "night" 8 hours or the least favourable full hour in the time between 10 p.m. and 6 a.m.
The European Environmental Noise Directive uses L_{den} (day-evening-night noise indicator) as its parameter, a deviation from the 24-hour assessment level. This parameter includes the long-term average sound level for the day (12 hours), the evening (4 hours) and the night (8 hours).

Peak level criteria
In addition to the assessment level, there are also requirements regarding the peak

T3: Recommended immissions values according to the Technical Rules for Protection Against Noise (*TA Lärm*) [dB(A)]

	Day	Night
Health resorts, hospitals and similar	45	35
Purely residential	50	35
General residential, small estates	55	40
Inner-city, villages and mixed uses	60	45
Commercial	65	50
Industrial	70	70
Within rooms requiring protection from sound irrespective of type of area	35	25

Individual, noise peaks may not exceed the recommended immissions values externally by more than 30 dB during the day and 20 dB during the night. In interiors, individual noise peaks may not exceed the recommended immissions values by more than 10 dB day or night.

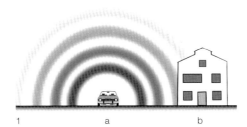

1   The sound emissions are characterised by the
    source and hence the cause, the immissions
    consider the effects.
    a Emissions point
    b Immissions point

levels that occur. Depending on the application, these lie up to 30 dB above the respective permissible assessment level.

Determination of noise exposure

The noise exposure can be determined by way of measurements or calculations. When assessing traffic noise loads, it is usual to apply a calculated noise level because the volume of traffic prevailing during measurements can vary considerably depending on time of year, day of the week and time of day.

Such calculations with subsequent assessments are carried out in two stages:

1. Determination of sound emissions, i.e. the noise radiated from a source or the power of the sound source.
2. Calculation of the sound immissions, i.e. the sound propagated from the source to the nearest buildings or areas requiring protection (Fig. 1).

Sound emissions calculation
In Germany the calculation of the sound emissions level, taking road traffic as the cause of the sound, is described in the Directive for Noise Abatement on Roads (RLS-90). The following factors contribute significantly to the generation of noise, i.e. sound emissions:

- Volume of traffic
- Traffic mix (proportion of HGVs)
- Permissible maximum speed
- Road surface

Methods of calculation for determining sound emissions exist for other noise sources as well, e.g. SCHALL 03 for rail traffic, VDI 3770 for sports.

Sound immissions calculation
When assessing new building work, the sound propagation calculation is carried out with the help of a computer, taking into account defined calculation rules. The propagation of noise is particularly affected by the distance between the source of the noise and the immissions point, man-made obstacles and reflective surfaces, and these factors must be considered.

The results of the sound propagation calculations can be output in the form of a table or a coloured noise map (building noise map or grid noise map).
Fig. 2 shows an example of a sound immissions calculation for a road. The noise exposure predicted by the calculations, i.e. the theoretical assessment level, is indicated in 5 dB steps.
These results can be used to decide whether additional noise control measures are necessary at the source or at some point along the transmission path. Acoustic requirements for the building facade, especially the windows, can also be derived from these results.

Measuring the noise exposure
Measurements to determine the noise exposure of existing commercial, industrial, sports and leisure facilities are generally carried out according to defined rules.
The measurement of sound immissions is understood to be the determination of the noise exposure direct at the point of the immissions.

If the sound immissions cannot be measured directly at the point where they occur because other noise sources act simultaneously, the critical sound sources are measured separately at the respective sources (measurement of sound emissions). Using the measurements as our starting point, the critical external noise level (sound immissions) can be determined with the help of a sound propagation calculation.
In a more accurate analysis of the critical noise producers, it is still advisable to

2 Urban planning design competition for the Bavarian
   Chamber of Insurers, Richard-Strauss-Strasse,
   Munich. Noise maps illustrate the external noise
   situation.
   Daytime assessment levels: a, b for an immissions
   point 11 m above ground level, and c, d in section.
   a, c Many apartments are exposed to noise
       because the existing buildings (1) lie perpen-
       dicular to the road.

b, d The planned transverse developments (2)
    create a quieter zone for the existing buildings.
    Suitable plan layouts in the new buildings allow
    the creation of additional attractive inner-city
    housing.

> 30.0 dB	> 50.0 dB	> 70.0 dB
> 35.0 dB	> 55.0 dB	> 75.0 dB
> 40.0 dB	> 60.0 dB	> 80.0 dB
> 45.0 dB	> 65.0 dB	> 85.0 dB

measure the noise levels of the individual
noise sources in situ.

Noise reduction at the source
Traffic noise
Limit emissions values for aircraft, trains
and road vehicles ensure an effective
reduction in the noise exposure.

Reducing the volume of traffic also helps
to reduce the noise level. Halving the
amount of traffic results in a 3 dB
decrease; greater reductions are possible
when the principal sources of noise, e.g.
HGVs, are diverted.
Speed limits also help to reduce noise.
With an HGV proportion of 10%, cutting
the speed limit from 70 km/h down to 50
reduces the level by 2 dB.

Another way of reducing the noise of road
traffic is to reduce the noise of the tyres
on the carriageway (rolling noise). For
example, at speeds of 50 km/h and more,
it is the rolling noise that dominates in the
case of cars.

Low-noise asphalt – a special open-
pore mix – reduces the noise level
permanently by at least 5 dB(A) compared to normal asphalt, which corresponds to a reduction in traffic amounting
to almost 70%. This achieves a noise
reduction over a wide area because a
lower noise exposure is to be expected
even in the higher storeys of apartment
blocks or buildings further away from
the road.
Following justified reservations with
regard to the laying and upkeep of such
open-pore asphalt mixes, this technology
underwent further development and is
now being used more and more, also in
urban areas. Particularly effective is a
two-coat open-pore asphalt whose coats
are adapted to suit the acoustic and
roadbuilding specifications.

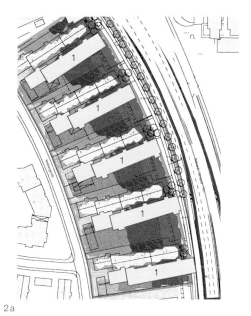

2a

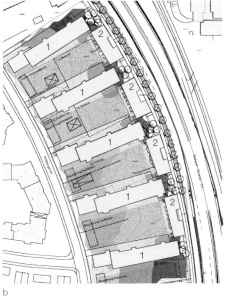

b

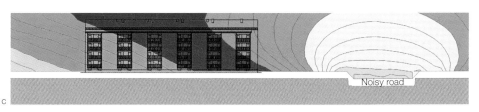

c

d

1a   b

Industrial noise quotas
According to the Technical Rules for Protection Against Noise (*TA Lärm*), the recommended immissions values must be applied to the sum of the sound immissions of all commercial facilities acting together at the immissions point. This means that in an industrial or commercial district, no further facilities can be approved once the recommended immissions values have already been reached or exceeded by the existing facilities. However, this does not apply when existing facilities have been closed down before starting up the new facilities or the sound emissions of existing facilities are reduced accordingly by way of voluntary or mandatory measures and the new facilities will not cause the immissions values to be exceeded.

But without corresponding precautions, it can happen that in an industrial or commercial district, the first facility (or one of the first facilities) exhausts the recommended immissions values and therefore effectively blocks the approval of any further operations or the expansion of existing operations.
To prevent this scenario, it is often the case these days that for those industrial and commercial districts with inadequate clearance to areas requiring protection from noise the development plan stipulates right from the start how much sound per square metre of area may be emitted so that the recommended immissions values in the surrounding area are not exceeded. Such emissions quotas can be specified either uniformly for an area or differentiated according to specific areas. DIN 45691 describes a noise quota method.

Methods for reducing industrial noise
Many sources of noise are in rooms or single-storey sheds. The ensuing noise levels in production shops and workrooms varies considerably from case to case. In order to limit the internal noise levels, it is usually best to attach a sound-absorbent material to the soffits. The noise escapes from the interior to the exterior mainly through openings (open windows, ventilation openings, unsealed joints, doors, etc.). Excessive noise propagation to the outside is avoided through the appropriate design and construction of the external components. In some cases it may be necessary to refrain from providing opening lights and install mechanical ventilation instead; doors and entrances may have to be built as airlocks to contain the noise. And sources of noise can also be found outside the building, e.g. fresh- and exhaust-air fans, exhaust-gas flues, works vehicles, etc. The provision of silencers, enclosures, etc. can contribute to reducing the noise level right at its source.

Noise due to sports, leisure and neighbourhood activities
More and more frequently, noise abatement also means controlling noise connected with leisure activities. The Sports Facilities Noise Abatement Act (18th BImSchV) contains measures that should be considered, especially when planning new buildings:

- Technical measures for loudspeaker systems (e.g. decentralised location, fitting of sound level limiters)
- Technical and constructional noise control measures at sports facilities (e.g. acoustically favourable floor coverings, low-noise ball barriers, noise control embankments/walls)
- Organisational measures so that spectators cannot use excessively noisy implements

c

d

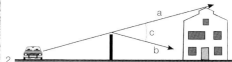

2

• Operational and organisational measures for the acoustically favourable design of routes to and from the facilities and parking areas

If the recommended immissions values are exceeded, the authority responsible can also specify opening times in order to guarantee that the values are adhered to. In doing so, however, protecting the neighbourhood and the general public must be weighed up against guaranteeing the sensible pursuit of sports activities.

Noise reduction in the propagation path
Increasing the distance
Increasing the distance between source and receiver, i.e. just moving the development further away from the source of the noise, can achieve a marked reduction in the level.
With an omnidirectional, point-like noise sound source (e.g. roof-mounted ventilation unit), the geometrical level decrease is 6 dB for each doubling of the distance. With linear sources such as roads and railway lines, the geometrical level decreases by only 3 dB when the distance is doubled.

Screening measures
Placing obstacles in the sound propagation path can achieve a noticeable reduction in the level. Familiar examples of this are the various forms of noise barriers alongside roads.

However, the noise-reducing effect of such noise barriers is typically limited to 5–10 dB because the sound is always diffracted around the edges of the barrier. The effect increases with the size of the diffraction angle and hence the deviation of the sound from its unobstructed propagation (without barrier). This means that the effect of the screening measure increases with the height of the barrier

and the closer it is to the source and the buildings needing protection. As the distances increase, so the effect decreases. Higher frequencies are better screened than low ones because the latter have long wavelengths that can be diffracted around the obstacle.

Where there is, for example, a terrace of houses behind the noise barrier, the barrier should be built from a sound-absorbent material in order to prevent reflections causing a rise in the noise level.

Examples of noise barriers
Noise barriers are mainly constructed in concrete, aluminium and timber. However, there are other materials that can be used in the design, e.g. glass, acrylic sheet and also special gabions (wire cages filled with stones). It should be ensured that the barriers exhibit adequate sound-damping properties so that the proportion of sound passing through the barrier is negligible compared to that diffracted around it. The sound insulation of noise barriers for the direct passage of sound should be at least 25 dB.
Planting a hedge or a single row of trees to protect against sound has proved to be ineffective, although the presence of a visual barrier can have a positive emotional effect. And despite their great mass, even gabions normally require sound-insulating layers because otherwise the sound "whistles" through the gaps between the stones!

Requirements regarding the sound insulation and sound absorption of noise barriers, e.g. when building noise barriers adjacent to trunk roads, can be found in publication ZTV-Lsw 06 (additional technical contractual conditions and guidelines for the construction of noise barriers for roads). The design of a noise barrier is regulated by both acoustic and urban planning

requirements. Noise barriers several metres high are not unusual.
However, for tall barriers a combination of embankment plus wall is to be recommended. It is also important to ensure that barriers are sufficiently long – in order to minimise the amount of sound that propagates around the ends of the barrier and into the area requiring protection.

Enclosures
One exceedingly effective, but also expensive, measure is to place the traffic route in a tunnel or some other form of complete enclosure. Even cuttings, to lower the level of the noise source, can prove effective.

Noise control at the building
If the aforementioned active measures are ineffective or inadequate, i.e. the level at the facade is higher than the desired or required level, it should be ensured that at least within the building, in the rooms where noise could be annoying or disturbing, there is no significant exposure to noise from outside.

Possible measures here include ensuring an acoustically favourable plan layout, the provision of suitable, acoustic facades (e.g. sound-insulating windows) or outer leaf constructions.

1   Various forms of noise barrier
    a Timber
    b Perforated sheet metal
    c Aerated concrete
    d Gabions
2   The screening effect of a noise barrier increases with the directness of the interruption to the sound propagation path, i.e. as the diffraction angle increases.
    Typical level reductions lie between 3 and 8 dB.
    a Direct sound
    b Diffracted sound
    c Diffraction angle

1

Plan layout

The sound level in front of the windows on the side of the building facing away from the source of the noise is about 5–10 dB lower than that on the side facing the source. One suitable noise control measure is therefore a favourable interior layout. Wherever possible, bedrooms and living rooms should be placed on the side facing away from the noise. Bedrooms and children's rooms in particular should be located where the assessment level L_r does not exceed 45–50 dB(A) during the night.

Furthermore, habitable rooms requiring protection from noise should be arranged in such a way that they can be ventilated with no or at least only minimal exposure to noise.

Transverse developments

Apart from that, transverse developments, e.g. buildings including commercial units with adjoining residential buildings behind, can offer an effective barrier against noise. Closing up gaps with partly glazed sur-faces is another option for reducing noise exposure (p. 43, Fig. 2).

Windows

If an assessment level of 58 dB(A) is reached or exceeded during the day, e.g. in front of habitable rooms or classrooms, it is necessary to verify that the external components of the building provide adequate sound insulation. The design of sound insulation for external building components for a given external noise exposure is carried out according to DIN 4109 "Sound insulation in buildings" or, for more accuracy, according to VDI 2719 "Sound insulation of windows and their auxiliary equipment" in conjunction with the 24th BImSchV.

The sound insulation required for the entire external surface of the room concerned is calculated depending on the assessment level in front of the facade and taking into account the room geometry and the frequency characteristic of the type of noise involved. From this it is

2

 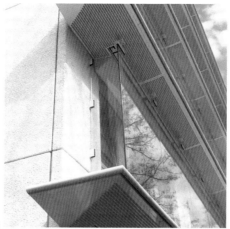

3a   b

then possible to calculate the weighted sound reduction index required for the windows by considering the proportion of window area in the facade plus the sound reduction indexes of the other parts of the facade.

The development of passive noise control measures for existing trunk roads in Germany is covered by a special directive (VLärmSchR 97), for railways by the document "Akustik 23".

To simplify the designation and selection of windows, they are divided into six sound insulation classes with a 5 dB difference in the sound reduction index between each class (see Tab. T5, p. 36). These days, sound insulation class 3 is achieved by many insulating glass units, but the costs increase noticeably from class 4 onwards. In many instances it is advisable to specify the windows not only according to sound insulation class, but also explicitly according to their weighted sound reduction index R_w. In addition, the spectrum adaptation term C_{Tr} must be included in order to define the sound insulation of the windows at lower frequencies as well. Further details and information concerning typical window assemblies can be found in the "Building acoustics" chapter (pp. 35–36).

Windows with good sound insulation values can reduce the level in the interior to such an extent that the problem of sleep being disrupted by traffic noise can be essentially avoided. But as such windows are only effective when closed, problems can arise due to the lack of adequate ventilation in the bedroom. Special attention must therefore be given to ensuring an adequate air change rate where sound-insulating facade constructions are used, which calls for mechanical ventilation or wall-mounted, low-noise ventilation fans.

*Natural ventilation in noisy environments –
(partly) glazed outer leaf constructions*
(Partly) glazed outer leaf constructions with an intervening cavity can be built in front of windows needed for the ventilation of habitable rooms. From the acoustic viewpoint, single glazing is adequate for such outer leaf constructions.
In the ideal case with a partly open window, the level in the interior can be reduced by a further 10–15 dB, which enables communication in the interior even when the external noise level is extreme.
When designing such outer leaf constructions or double-leaf facades, there is always a conflict between the acoustic and interior climate requirements. In order that the intervening cavity does not heat up excessively, the air must be able to circulate. The openings required for this in turn reduce the sound insulation.
Sound-absorbent linings to the window reveals or on the soffits of outer leaf constructions can improve the sound insulation. With a double-leaf facade there is also the risk of flanking transmissions from one room to another.
This aspect is investigated in more detail in the chapter entitled "Office buildings".

3   Close-up views of facade, office building refurbishment, Berg-am-Laim-Strase, Munich, 2000, Angerer Demmel Hadler Architekten
    a   Various options were evaluated acoustically on a typical facade. This enabled the best possible solution to be found for the conflicting demands of sound insulation and interior climate.
    b   After completion: acoustically adapted ventilation slits and an absorbent design to the underside of the cantilevering panels ensure that communication indoors is possible with the windows partly open – even in the case of extreme external noise levels.

Sound insulation as a quality feature in housing

Noise control is especially important in housing because it plays a great role in the health and well-being of people. Our living accommodation should offer us the chance to withdraw, should provide a private sphere, which calls for adequate acoustic screening from our neighbours. In this respect, sound insulation is relatively easy to define from the physical point of view. But from the point of view of users' emotions, identical physical conditions inevitably give rise to large discrepancies in how users assess those conditions. And users' expectations quite rightly depend on the location, furnishings and fittings and the price of a property. Consequently, the aspect of sound insulation must be considered carefully during design and construction and, in the end, presented as transparently as possible for clients or buyers.

Sound insulation categories and legal situation

Minimum sound insulation to DIN 4109
In Germany DIN 4109 "Sound insulation in buildings" is the most important noise control standard included in building legislation, especially for housing.
The DIN standard distinguishes between, on the one hand, multi-storey buildings containing apartments and workrooms and, on the other, semi-detached and terraced houses. As can be seen from Tab. T2 (p. 51) and Tab. T3 (p. 53), low-rise houses are assigned a higher passive noise control level than the apartments in multi-storey buildings.
However, DIN 4109 only specifies the minimum legal building acoustics requirements laid down in public law, which must always be complied with. Specifying a minimum standard only provides protection against unreasonable annoyances; users should in no way expect that noises from neighbouring apartments or from outside can no longer be heard. This

results in the need for mutual consideration. Sound insulation to DIN 4109 is therefore not a quality seal, but rather an absolute necessity in line with our cultural customs.

Enhanced sound insulation to DIN 4109 supplement 2
DIN 4109 supplement 2 provides recommendations for better sound insulation. Complying with these recommendations will result in the majority of occupants being satisfied with the level of sound insulation without having to adjust their lifestyle to any significant extent. In multi-storey housing, loud voices in a neighbouring apartment may well be audible, but generally unintelligible. The playing of musical instruments or singing will, however, remain audible in neighbouring apartments.

Acoustic disturbances are generally ruled out in semi-detached and terraced houses with enhanced sound insulation to DIN 4109, even if, for example, a piano is being played next door.

Sound insulation categories to VDI 4100
VDI 4100 "Noise control in dwellings – Criteria for planning and assessment" specifies differentiated requirements for three acoustic quality grades. Here again, a distinction is made between multi-storey

apartments and semi-detached and terraced houses. Tab. T1 shows a subjective classification of the different sound insulation levels for multi-storey apartments. VDI 4100 sound insulation category I is identical with the minimum requirements of DIN 4109; sound insulation category II for multi-storey apartments lies approximately on the level of better sound insulation meeting the enhanced requirements of DIN 4109 supplement 2; and sound insulation category III for terraced houses corresponds approximately to DIN 4109 supplement 2. There were efforts to combine the contents of VDI 4100 and DIN 4109 supplement 2 in an effort to provide unified recommendations.
The outcome was a draft standard, DIN 4109-10 "Proposals for enhanced sound insulation in dwellings", published in 2000, in which a three-level sound insulation concept was proposed. But the building and housing industries opposed this vehemently because they feared that a three-level system would generate further legal uncertainties, an increased risk of liability and higher building costs. In the end, the objections led to the withdrawal of the draft standard in 2005. As a consequence of this, August 2007 saw the publication of an updated version of VDI 4100. Planners and architects involved with housing therefore still have the choice of using DIN 4109 supplement 2

T1: Awareness of typical noises from neighbouring apartments and their allocation to three sound insulation categories (SSt) for a background evening noise level of 20 dB(A) to VDI 4100

Noise emission	SSt I[1]	SSt II	SSt III
Loud speech	intelligible	generally intelligible	generally unintelligible
Normal speech	generally unintelligible	unintelligible	inaudible
Walking, footsteps	generally disturbing	generally no longer disturbing	not disturbing
Noises from building services	unreasonable annoyances generally avoided	occasionally disturbing	not or rarely disturbing
Playing music, loud radio/ television	clearly audible	clearly audible	generally audible

[1] Corresponds to minimum sound insulation to DIN 4109

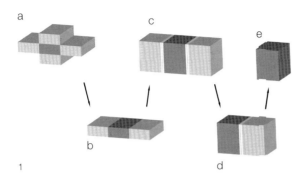

1   Acoustic ranking of different building types
    a Multi-occupancy building
    b "Terraced apartments"
    c Terraced houses
    d Semi-detached houses
    e Detached house
2   In principle, it is possible to demonstrate relation-
    ships between the sound insulation achievable
    and typical forms of construction.
    Schematic horizontal sections:
    a Lightweight clay brickwork
    b Heavyweight clay brickwork
    c Concrete box-frame construction

or VDI 4100 when enhanced sound insulation is required.

Building types

As described above, semi-detached and terraced houses have to satisfy higher passive noise control requirements than multi-storey apartments. In addition, the type of building, from multi-occupancy block to detached family home, has an effect with respect to the potential sources of noise, i.e. the number of neighbours. With fewer neighbours, the degree of disturbing noise to be expected, and also the consideration that must be shown towards neighbours, is also lower. Fig. 1 shows common building types in their acoustic order.

In a multi-occupancy block, an apartment can adjoin up to five others in the most unfavourable case. On the other hand, it is possible to plan multi-occupancy blocks so that the number of immediate neighbours is minimised, as is the case with rows of apartments separated by a corridor and, in particular, penthouses. Building forms like "townhouses" or "quattro houses" (= four dwellings in a square building) at least avoid the situation where different housing units are on top of each other. However, such buildings often employ the construction principles of multi-occupancy blocks, which then result in the sound insulation not achieving the standard that was perhaps expected.

Compared to terraced houses, the acoustic quality of semi-detached houses is improved only by the fact that there is one neighbour instead of two.

With detached family homes, the acoustic quality is defined via the distance to the nearest neighbour, possibly also the sound insulation within the building itself.

Legal uncertainties

It is not uncommon for clients or users to be so dissatisfied with the sound insulation of their property that the dispute has to be settled in court. One problem is that the contents of building contracts are sometimes ambiguous, possibly even contradictory, when it comes to describing the sound insulation. In such situations it is not, for example, the minimum requirements of DIN 4109 that are relevant, but instead civil law, which can demand considerably higher sound insulation values. For instance, a verdict from the German Federal Court of Justice dating from 1998 (ref. No. VII ZR 184/97) states: "Which sound insulation has to be provided is to be determined from the meaning of the contract. If certain sound reduction indexes have been expressly agreed in the contract, or in any case are to be achieved through the meaning of the contract, the execution of the works is deficient if these values are not achieved. If such an agreement is not available, the execution of the works is generally deficient if the works do not correspond to the acknowledged technical standards as a contractual minimum standard at the time of acceptance."

The consequence of this legal practice, also confirmed by further judgments (e.g. Federal Court of Justice verdict of 14 June 2007, ref. No. VII ZR 45/06), is that – where unambiguous contractual provisions are lacking – the sound insulation to be normally achieved with the form of construction is to be provided, or an acoustic quality specified by the court in conjunction with other qualities. In some circumstances this quality may be much higher than the minimum requirements of DIN 4109, which in the end is determined by a specialist appointed by the court.

In addition, established forms of construction and the sound insulation achievable with those forms are assessed as generally acknowledged technical standards. The best example of this is the double-leaf party wall between terraced or semi-detached houses, which normally achieves a far higher sound reduction index than the minimum value given in DIN 4109.

These judgments also show, however, that the contractual agreements regarding certain properties, i.e. in the end the sound reduction index, the normalised impact sound level or the sounds from building services, assume maximum priority in legal disputes. Architects, clients and contractors would therefore be well advised to include such requirements quantitatively and explicitly in their contracts following a detailed analysis of the situation. This must be explained to the client in detail in advance, especially when there is a noticeable discrepancy between the building specification and the passive noise control. This is the case, for example, when the specification contains phrases like "high-quality fitting-out", but the sound insulation is merely referred to as "in accordance with the DIN standard". Legal uncertainties sometimes also result from the aforementioned "townhouses" or "quattro houses", which from their very nature lead us to expect a higher standard of sound insulation than is the case, for example, with a traditional multi-occupancy apartment building.

The acoustic planning and monitoring process

In the light of the aforementioned aspects concerning the type of construction required, the aims of noise control within a building should be defined together with the client at the preliminary design stage. The sound insulation specification can be based on the VDI 4100 sound insulation categories described above or

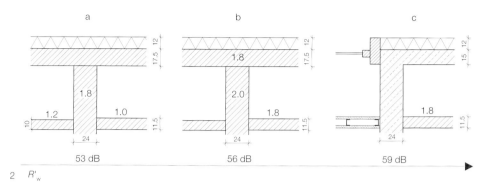

2 R'_w

T2: Minimum requirements to DIN 4109 and parameters for sound insulation categories (SSt) for apartments in multi-occupancy buildings to VDI 4100 (excerpt)

		Variable	DIN 4109[1]	SSt II[2]	SSt III
Airborne sound insulation	between habitable rooms and other dwellings	horizontal vertical	R'_w [dB]	53 54	56 57	59 60
	between habitable rooms and stairs/halls not in the same dwelling			52	56	59
Impact sound insulation	between habitable rooms and other dwellings	$L'_{n,w}$ [dB]	53	46	39
	between habitable rooms and stairs not in the same dwelling			58	53	46

[1] Corresponds to VDI 4100 category SSt I
[2] Sound insulation level comparable with the recommendations for enhanced sound insulation to DIN 4109 supplement 2

DIN 4109 or DIN 4109 supplement 2. When building apartments for sale, these aims must be coordinated with the other objectives of the construction project. For high-quality housing units there is also the chance of defining quality targets within the dwellings.

The achievement of these aims, taking into account the plan layout and the sound insulation values generally achieved with the building materials to be used, must be checked within the scope of the design work. This planning process should be set up as a control loop, in terms of both choice of building materials and costs.
An appropriate, professional analysis of the sound insulation should be carried out at the end of the design work. This is currently carried out according to DIN 4109 supplement 1 and in future according to rules based on Euronorms.
As proof of the quality of workmanship, these defined targets should be checked through quality certification of the sound insulation, by measurements for individual propagation paths.
These planning and monitoring processes can be used to define and verify individual qualities, as well as an overall quality if necessary. In particular, verification by way of measurements by an acoustics

institute should be accorded an appropriate priority in the overall assessment.

Multi-storey apartment buildings
As described in the foregoing, design and construction are frequently carried out based on the minimum requirements of DIN 4109. Higher requirements can be agreed upon in private contracts, and DIN 4109 supplement 2 or VDI 4100 are suitable benchmarks in such cases. Of course, other, more stringent requirements can also be specified.

Typical forms of construction
Fig. 2 illustrates a sensible method for distinguishing between typical forms of construction for multi-storey apartment buildings from the acoustics viewpoint. Building systems, e.g. monolithic external walls built from lightweight vertically perforated clay bricks with good thermal insulation properties, can also be integrated into this approach.
Qualities between lightweight and heavyweight clay masonry are conceivable for these materials. In particular, the concrete box-frame construction shown, coupled with prefabricated facades and dry construction internally if necessary, can be expected to achieve the highest possible acoustic quality in multi-occupancy housing, even managing VDI 4100 category SSt III (e.g. for party walls: $R'_w \geq 59$ dB).

Party walls
In principle, the typical forms of construction used for multi-storey apartments can all achieve the minimum sound insulation to DIN 4109 (corresponds to VDI 4100 category SSt I).

For example, according to DIN 4109 supplement 1, a mass per unit area of 410 kg/m[2] is necessary for a single-leaf solid party wall in order to guarantee a value of $R'_w \geq 53$ dB. One way of achieving this is to use a 240 mm thick masonry wall (clay bricks, density class 1.6 or higher) plastered both sides. Theoretically, the sound insulation requirement is properly fulfilled when the average mass per unit area (arithmetic mean) of flanking solid components is at least 250 kg/m[2]. This is also achieved by plastered and rendered external walls 360 mm thick made from vertically perforated clay bricks of density class 0.7, although the negative effect of resonance within the clay bricks should be taken into account (see p. 29, "Wall constructions").

Notwithstanding, this form of construction cannot be analysed with the methods currently still valid. Such forms of construction should therefore be regarded as building systems. Besides the mass per unit area m', it is also important to consider certain types of joints, connections and junctions (floors, walls) and the flanking lengths both sides of the separating element. The manufacturer must be involved in the verification procedure and must specify the details required.
Where higher sound insulation requirements are necessary, forms of construction employing heavyweight clay bricks or concrete are certainly an advantage, although satisfying the enhanced sound insulation requirements of DIN 4109 supplement 2 (VDI 4100 category SSt II), e.g. with vertically perforated clay bricks as

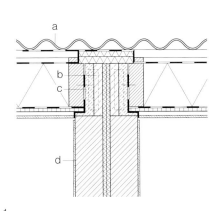

1

2

1 An acoustically optimised junction at the roof is necessary in order to avoid reducing the sound insulation quality of a double-leaf party wall in the roof space.
   Roof construction:
   a Corrugated fibre-cement sheeting, natural colour
      40 x 60 mm battens
      24 x 48 mm counter battens
      waterproof sheeting
      240 mm insulation between
      60 x 240 mm rafters
      vapour barrier
      15 mm OSB as stiffening diaphragm
      24 x 48 mm battens
      12.5 mm gypsum fibreboard
   b In situ concrete filling cast after erecting roof
   c Precast concrete element
   d 150 mm calcium silicate masonry
2 Terrace houses in Göppingen, 1999, Wick + Partner
   The separating joint between the houses is clearly visible.

flanking components, need not be ruled out in principle.

In order to achieve the enhanced sound insulation according to DIN 4109 supplement 2, i.e. $R'_w \geq 55$ dB, a wall with a weight per unit area of 490 kg/m² is required, for instance, which can be realised in practice with a 240 mm thick, plastered masonry wall with clay bricks of density class 2.0. Again, this statement is only valid when the flanking components exhibit an average weight per unit area of at least 350 kg/m². In the case of flanking external walls made from vertically perforated clay bricks, protection against flanking transmissions must be given special attention when attempting to satisfy the requirements for enhanced sound insulation.

Party walls with a high mass per unit area can also be achieved with hollow clay bricks that are filled with concrete on site. However, all voids within the bricks must be filled completely because otherwise resonance effects can reduce the sound insulation.
And finally, from the acoustics viewpoint it is also possible to build party walls using dry construction techniques.

Party floors
Beside the depth of the structural floor, which is determined by the structural calculations, the choice of insulating material and the thickness of the screed also result in a large variation, as was demonstrated before in the case of airborne sound insulation (see p. 33, "Suspended floors and floor finishes"). The enhanced acoustic requirements of DIN 4109 supplement 2 regarding airborne and impact sound insulation can usually be met by typical forms of construction, e.g. a 160 mm deep reinforced concrete slab plus an acoustically favourable floating screed with an impact sound reduction index $\Delta L = 28$ dB (p. 33, Tab. T3).

It should be remembered that according to DIN 4109, resilient floor coverings, e.g. carpeting, or real wood laminate flooring on a layer of insulation, may not be included in the calculations; so floating screeds are the norm in the construction of multi-storey apartments.

Staircases
In new buildings it is very often the transmission of impact sound from staircases to adjoining apartments that leads to complaints. The causes are poorly separated screeds at the entrances to the individual apartments, or reinforced concrete stair flights connected rigidly to the walls, either due to a flaw in the design or incorrectly laid finishes.

Apartment entrance doors
Where apartment entrance doors lead to a hallway serving rooms requiring protection against noise (separated by doors and especially living and working areas or bedrooms), the entrance door must achieve a weighted sound reduction index of $R_w \geq 27$ dB. However, where the apartment entrance door opens directly into a room requiring protection, a much higher value of $R_w \geq 37$ dB is required, necessitating a much more elaborate, more expensive door design.

Roof space conversions
In existing buildings, but also in new buildings, the roof space is often converted to create attractive living spaces. Characteristic of this conversion work, and at the same time a weakness in terms of the acoustics, is often the junction between a party wall and the underside of the roof, and the flanking transmissions via the roof. This detail must therefore be given due attention (Fig. 1).

Refurbishment projects
When refurbishing buildings, the owner can expect that the building measures correspond to the generally acknowledged technical standards for sound insulation as well, i.e. comply with the minimum requirements of DIN 4109 (latest edition), perhaps even exceed these. The design of such measures is usually carried out according to the individual circumstances following a survey that takes into account the framework conditions such as structural requirements, depth of intermediate floors plus finishes and economic viability. Exceptions may be necessary in order to satisfy the stipulations of preservation orders.

Building services
When considering the propagation of noise in the form of structure-borne sound, e.g. from water pipes or other installations, the number one rule remains valid: a favourable plan layout. Good sound insulation normally requires some form of buffer space between the source of the sound and rooms requiring protection.
The acoustic quality of the installations is generally very much dependent on the particular systems. In particular, placing water pipes behind a wall lining in dry construction is almost unavoidable these days when a reasonable level of sound insulation is required.

Semi-detached and terraced houses
DIN 4109 contains the basic requirements for minimum sound insulation to be considered when planning semi-detached and terraced houses. Again, higher standards are described in DIN 4109 supplement 2 and VDI 4100. The configuration of the party wall between the houses is critical for the sound insulation.

T3: Minimum requirements to DIN 4109 and parameters for sound insulation categories (SSt) for semi-detached and terraced houses to VDI 4100 (excerpt)

		Variable	DIN 4109[1]	SSt II	SSt III[2]
Airborne sound insulation	between habitable rooms and other dwellings	R'_w [dB]	57	63	68
Impact sound insulation	between habitable rooms and other dwellings	$L'_{n,w}$ [dB]	48	41	34
	between habitable rooms and stairs/landings not in the same dwelling		53	46	39

[1] Corresponds to VDI 4100 category SSt I
[2] Sound insulation level comparable with the recommendations for enhanced sound insulation to DIN 4109 supplement 2

Semi-detached and terraced houses are normally built with separate party walls, i.e. a double-leaf construction. This solution enables a high level of sound insulation to be achieved which complies with enhanced noise control requirements.

The factors that influence the acoustic parameters are, in particular, the mass per unit area of the individual leaves and the width of the cavity. The heavier the leaves and the wider the cavity, the better is the sound insulation. Doubling the width of the cavity has the same effect as doubling the mass per unit area of the two leaves.
DIN 4109 supplement 1 specifies a minimum cavity width of 30 mm, but wider cavities of 60–80 mm are to be recommended from the acoustics viewpoint because this solution increases the sound insulation markedly without any significant extra cost or extra work or loss of floor space. In order to attenuate the cavity and minimise the risk of bridges for structure-borne sound, e.g. due to lumps of

mortar, mineral insulation 30–40 mm thick should be affixed to the whole area of the leaf built first, e.g. with a thin layer of mortar. Nails should not be used because they could form acoustic bridges. This thickness of insulation is adequate even for wider cavities.
Heavyweight clay brickwork (e.g. density 1800 kg/m³) 115 mm thick, but also thicker, lighter clay bricks or precast concrete elements, can be used for the wall leaves. Double-leaf in situ concrete walls are also possible, but special rigid cavity insulation is then required. Owing to the higher stiffness of such insulation, a double-leaf in situ concrete wall is – in acoustic terms – less effective than a wall of the same thickness made from precast concrete elements.
It is vital to ensure that no bridges for structure-borne sound ensue at the junctions with the intermediate floors. This means that formwork is needed for concrete floor slabs adjacent to the separating cavity. The external walls, too, must be separated at the party wall – also walls

below ground if applicable. Only when the basement walls are built with impermeable reinforced concrete because of hydrostatic pressure is the joint in the basement walls freely omitted.

The factors described here influence the sound insulation that can be achieved in the end. Surveys have revealed that the use of properly constructed double-leaf walls with a basement enable values of R'_w = 65 dB to be reached at ground floor level and R'_w = 62 dB in buildings without a basement. Comparable values are also possible when the aim is to reflect the generally acknowledged technical standards. But in fact, as mentioned above, the aims of noise control should be explicitly defined and the party wall designed accordingly to suit each individual situation.

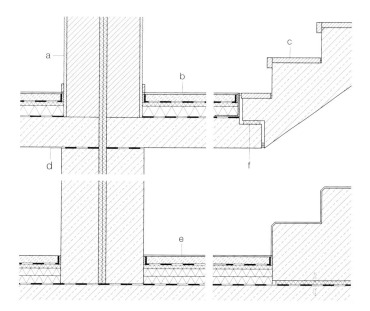

3

3 If a double-leaf party wall extends into the basement, it is possible to achieve weighted sound reduction indexes of R'_w ≥ 65 dB at ground floor level. Floating screeds and stairs bearing on resilient pads ensure the desired level of noise control both for neighbours and within the house itself.
   a Wall construction:
     plaster, calcium silicate masonry, 175 mm
     insulation, 2 No. 20 mm
     calcium silicate masonry, plaster, 175 mm
   b Floor construction (ground/upper floors):
     8 mm oak wood-block flooring, levelling layer
     40 mm screed, PE sheet as separating layer
     20 mm impact sound insulation
     60 mm rigid foam thermal insulation
     PE sheet, 160 mm reinforced concrete,
   c Stairs:
     27 mm oak laminboard tread
     25 x 50 mm oak nosing
     180 mm precast concrete element
   d Separating layer
   e Basement floor construction:
     5 mm carpet, 40 mm screed
     PE sheet as separating layer
     95 mm rigid foam thermal insulation,
     waterproofing, 250 mm impermeable reinforced concrete
   f Resilient bearing pad

Sound insulation as a quality feature in housing
Sound insulation within the dwelling
DIN 4109 – history and future planning instrument

1   Development of minimum acoustic requirements
    for party walls in housing

If the party wall is not to be built as a double-leaf heavyweight wall, it is certainly still possible to achieve the minimum sound insulation according to DIN 4109 even with a single-leaf wall. In such a case, however, it is imperative to inform the client or buyer about this in detail because the level of sound insulation achieved is far lower than that for the forms of construction normally employed these days in Germany.

Sound insulation within the dwelling

Noise control objectives can also be formulated for the areas within an apartment or semi-detached, terraced or detached house. These are always additional agreements included in private contracts; no building legislation applies here.

For example, the recommendations regarding normal sound insulation within one housing unit according to DIN 4109 supplement 2 can generally be implemented without significant extra costs or extra work. However, in timber houses or when using timber joist floors where the joists are to remain visible, the desired level of sound insulation and the associated building measures should be planned at an early stage in order to avoid the disappointments of poor sound insulation later.

In the really quite narrow terraced houses frequently encountered in urban areas these days, the stairs are usually directly adjacent to the living areas. But this more generous use of the interior space means that the sound can easily propagate to the upper floors. If acoustic separation between living areas and, for example, bedrooms or children's rooms is important, the different areas should always be separated by two doors.

In an open-plan layout with a combined living, dining and kitchen area, less than optimum acoustic conditions are sometimes apparent. The larger the volume of the room, the longer is the reverberation time, which reduces the legibility of speech – also that emanating from a television – and makes other noises, e.g. a dishwasher in the kitchen, more readily noticeable. In rooms with a normal ceiling height and a floor area less than about 40–50 m², typical furnishings and fittings, e.g. sofa, chairs, bookshelves, possibly curtains and separate rugs, usually solve this problem. But in the case of larger rooms and/or higher expectations regarding the room acoustics quality, e.g. for high-end audio or home cinema equipment, the use of building measures to reduce the reverberation time should certainly be considered.

External noise situation

General statements regarding the acoustic exposure of a plot can be found in the land-use plans. Detailed information is shown on the local development plans. Finally, the building approval documents, possibly containing stipulations regarding noise control, enable the external noise situation to be appreciated in detail.
In particular, in "purely residential areas" (designation WR) and "general residential areas" (WA), it can be assumed that healthy sleep with partly open windows is possible when the recommended immissions values characteristic for the area are not exceeded. The allocation of the area, however, does not rule out the fact that on the boundaries, or in established structures, the recommended immissions values could be exceeded, which may result in stipulations regarding the construction. This is particularly true for central urban or mixed-use areas.
Noise barriers represent one possible solution, rows of garages another, both of which screen housing and recreational areas against noise. Measures on the buildings themselves include sound-insulating windows, glazed loggias, possibly low-noise wall-mounted ventilation fans; interior layouts and the building orientation can also be optimised to achieve the best acoustic situation.

DIN 4109 – history and future planning instrument

This chapter will be concluded with a brief excursion into the history of DIN 4109, but especially an overview of the changes to be expected which may well characterise the acoustic planning of housing in the future.

Development stages of DIN 4109
The first edition of DIN 4109 dates from the years 1938/1944. It showed the sound insulation required by means of the forms of construction typical in multi-occupancy housing at that time. The party wall in the form of 240 mm thick solid clay brickwork was a key element in the construction. Fig. 1 shows the development in the minimum requirements for party walls in residential buildings. Besides emphasizing the health aspect, this development can also be regarded as the retention of a building culture typical of Germany. For that time the sound insulation achievable, i.e. the sound reduction indexes attainable, can certainly be assigned the attribute "in accordance with the technical standards". This also applies to the solutions illustrated for achieving adequate impact sound insulation with carpeting or hard finishes on a separating layer. Looking back, combining the quality to be realised and forms of construction according to the technical standards seems to be sensible and appropriate.

However, developments in construction would seem to make this connection

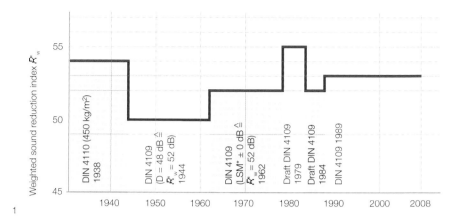

1

*LSM = airborne sound insulation margin

questionable. Both the further developments in the realm of the theoretical models for describing sound insulation in buildings and also the considerable expansion and variation in building materials and forms of construction permit a greater differentiation between typical forms of construction these days. The context of "building according to the technical standards" must be looked at more closely. Forms of building according to the technical standards can enable a wider area of acoustic qualities to be described.

By the time the 1962 edition of DIN 4109 appeared, the standard represented a self-contained system for creating the fundamentals for planning based on construction law requirements and recommendations for better sound insulation. By means of a calculation model based on measurements of test setups with flanking paths typical of real buildings, a high degree of planning reliability could be achieved in advance. A model for calculating the sound insulation in frame construction was introduced in addition to the model for solid, heavyweight construction (taking flanking components into account). The methods for checking components in test setups and for monitoring the quality of structures already built – described in subsequent supplements – rounded off this early quality management system for noise control in buildings.

The 1989 edition covered another level of detail which is still valid today. Increasing the reliability of workmanship when laying floating screeds, coupled with great variation in impact sound insulation materials represent the main progress at that time.

What can we expect in the future?
The current revision and new edition of DIN 4109 were rendered necessary by EU stipulations such as new measuring techniques for identifying building components and calculation rules.

The new edition of DIN 4109 will contain the following parts:
- Part 1: Minimum requirements for sound insulation
- Part 2: Calculations for verifying compliance with the requirements
- Part 3: Input data for the calculations for verifying the sound insulation (national building components catalogue)
- Part 4: Performance of building acoustics tests (safety concept)

In the course of these revisions, new descriptive variables, "parameters related to reverberation time", were proposed in the draft of DIN 4109, October 2006 edition. These are the weighted standardised sound level difference $D_{n,T,w}$, the standardised impact sound level $L'_{n,T,w}$ (see p. 27, "Designating the sound insulation of components") and levels related to the reverberation time for noises caused by building services and operations.
For airborne sound insulation the new requirements relate directly to the reduction in level to be expected between two rooms, for impact sound insulation directly to the impact sound level transmitted.
Besides separating sound insulation (expressed by way of level differences) and building technology requirements (sound reduction indexes R'_w etc.), the changeover provides the possibility for international comparisons. Most European countries already use the variables standardised sound level difference or standardised impact sound level.
As a consequence of this, requirements are no longer placed on separating components, but rather on the sound insulation between rooms used for different functions.

New planning process
The ongoing development of DIN 4109 with this notation results in the opportunity to specify requirements or definitions of desired acoustic qualities irrespective of the forms of construction. If desired, this can be extended downwards to room-by-room definitions. Detailed sound insulation concepts for noise control within buildings can be specified purely on the basis of the needs of the client/user.

On the other hand, the fear of reduced acoustic quality is fuelled because lower sound reduction indexes for achieving the minimum requirements are adequate for room depths exceeding 3 m and last but not least, the noise control planning becomes more involved because this approach can result in different wall constructions being specified to achieve the same protective targets depending on the size of the room and the area of the separating wall. Hitherto this was not the case because it was solely the properties of the building components that were critical.

As mentioned above, fears that building would become more expensive, particularly in the housing sector, has led to minimum sound insulation requirements being included in the new DIN 4109 as well. Harmonisation with standards and directives that describe higher acoustic qualities could not be realised. Consequently, several standards and directives will continue to remain in circulation for the foreseeable future.

Inner-city development, Munich

Architects: Fink + Jocher, Munich
Acoustics
consultants: Müller-BBM, Planegg
Completed: 2005

Westend is an aspiring district not far from the centre of Munich. This inner-city development replaces five older buildings whose upkeep was no longer proving economic. With its commercial/housing mix, this development helps to upgrade the district. The properties offer the possibility modern living and working conditions within the city.

The structural elements of the building have been reduced to the external walls and staircase cores. The entire fitting-out with lightweight walls means that the interior can be adapted to the needs of users throughout the lifetime of the building.

All rooms benefit from underfloor heating and controlled ventilation. There is a very busy main road nearby and so the fresh air is drawn in via the roof. A heat exchanger minimises the ventilation heat losses.

The excellent thermal insulation properties of the external components mean that the building remains within the limits for a low-energy building.

The adjoining main road is used by 32 000 vehicles every day plus two tram lines, which produce sound emissions of up to 75 dB. This noise exposure is not unusual for inner-city locations.

In contrast to the strategy often used, in this inner-city development it is not only ancillary rooms with small windows that face onto the road. Coupled windows with good sound insulation – double glazing on the inside, single glazing on the outside and sound-absorbing surfaces in the intervening cavity – enable living spaces to be placed on the road side as well. The orientation of the apartments is not restricted, which provides opportunities to exploit the incident solar radiation. In addition, the road becomes a public, socially controlled external space.

1   Commercial usage
2   Communal space
3   Maisonette
4   Loggia
5   Rooftop terrace
6   External wall construction:
    20 mm render (with glass splinters on road side)
    140 mm thermal insulation
    200 mm reinforced concrete
7   Single glazing,
    8 mm toughened safety glass
8   Wooden window frame with double glazing, sound insulation according to calculations
9   Splayed window reveals lined with 2 mm factory-bent perforated sheet aluminium
10 Sound insulation, 30 mm impact sound insulation board
11 Thermal insulation, 40 mm
12 Window head lined with 3 mm factory-bent perforated sheet steel
13 Square aluminium tube with rubber gasket, 25 × 25 × 3 mm
14 Thermal insulation, 50 mm
15 Aluminium stay connecting outer and inner window frames, 25 × 10 mm
16 Rectangular aluminium tube, 100 × 40 × 4 mm
17 Steel angle with stiffeners, 160 × 225 × 4 mm
18 Aluminium angle, 25 × 25 × 4 mm
19 Aluminium angle, 70 × 30 × 5 mm
20 Fixed glazing as safety barrier, laminated safety glass (2 No. 8 mm float glass)

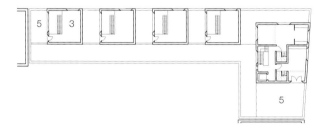

Rooftop storey

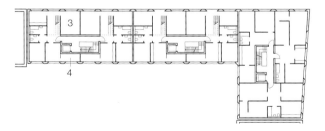

3rd floor

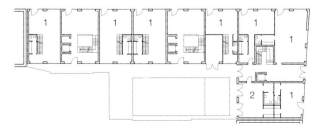

Ground floor

Plans
scale 1:750
Sections through coupled window
scale 1:10

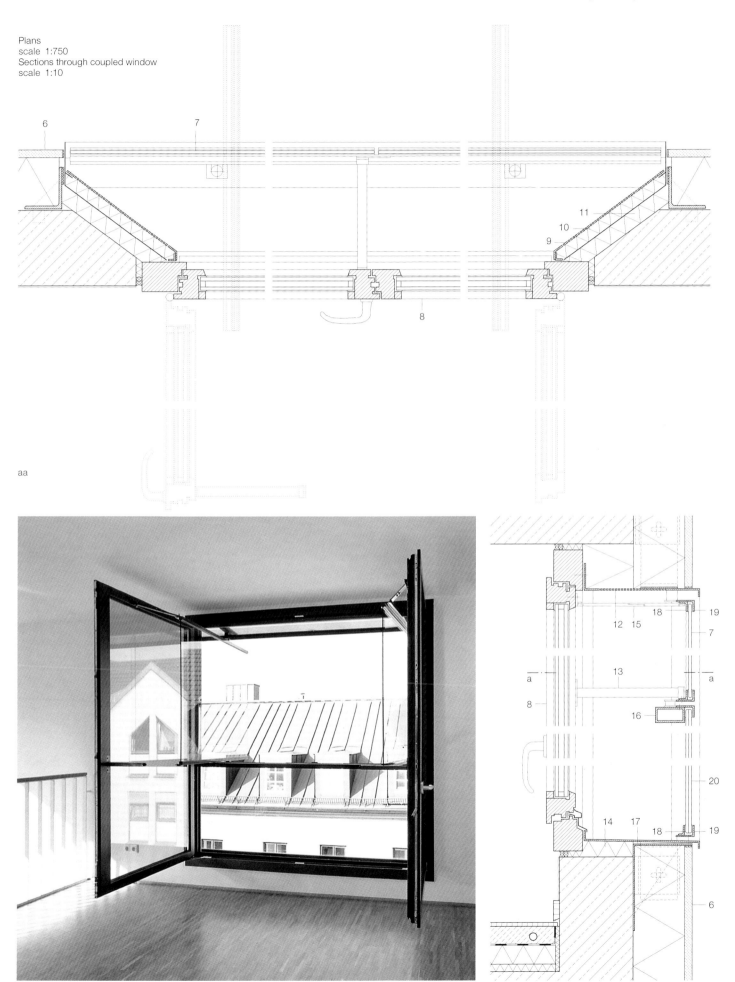

aa

Hardly any other type of building has to undergo such rapid changes as the modern office building now designed to accomodate adaptability, the abandonment of fixed room structures and non-territorial office structures. The fitting-out must be designed to react flexibly to future demands and to permit new interior layout concepts. The acoustic aspects in this changing structure are also complex, and there are frequent disappointments after a building is completed and occupied. It is especially disturbing when conversations are readily intelligible – whether they take place in the next office cubicle or the neighbouring workstation in an open-plan layout. Excessive reverberation – in an office cubicle, in an open-plan office, in public areas or in multi-functional atria – can also detract from the quality of an office building.

Open-plan office structures
Offices with an open-plan layout are very popular. The advantages frequently mentioned include the encouragement of informal communication and teamwork plus more efficient use of the floor space. But there are also disadvantages, some of which are of an acoustic nature. Expo-

sure to disturbing noises and/or the ease of intelligibility of (telephone) conversations conducted at neighbouring, even more distant, workplaces are among the complaints.

Such conflicts can be reduced in open-plan office structures, albeit without attaining the acoustic quality of separate offices. Conversations from nearby areas remain audible and readily intelligible to a greater or lesser degree depending on the layout of the area. This may be tolerable within work teams but is probably undesirable between different departments with different requirements.

What this means in the end is that the acoustic factors, too, should be carefully considered right from the beginning when specifying an office concept. Only those workplaces that actually have to communicate with each other should be combined into one office. The orientation of the workplaces, or rather the direction of communication, plays a role here, likewise the distance between neighbouring workplaces. If telephones are in frequent use, the use of headsets could be a viable solution.

1

2a b

1 Conversion of industrial building, KMS TEAM, Munich design office, 2007, Jürke Architekten
An open-plan office in the truest sense of the word! The acoustic screening in the longitudinal direction of this single-storey shed is provided by shelving. Large-format ceiling panels made from wood-wool acoustic boards between the steel rafters and the height of the room prevent disturbing reflections from the ceiling.
2 Constructional room acoustics measures in open-plan offices with thermally activated slabs.
 a Perforated sheet metal panels with an absorbent backing plus a sound-absorbent soffit to the ducts for services (perforated metal boxes)
 b Sound-absorbent baffles made from mineral-fibre insulating material

Communication zones, tea kitchens, also printers and photocopiers should be screened off from working areas or, even better, placed in separate rooms.

Furthermore, three acoustic criteria must be harmonised in open-plan offices: adequate acoustic attenuation in the room, a background noise level that is not too low, and effective screening.

Acoustic attenuation in the room
Appropriate acoustic attenuation, i.e. decreasing the reverberation, is necessary in open-plan offices. In accordance with the relevant standards and directives (DIN 18041, VDI 2569), this requirement is expressed in the form of the so-called A/V ratio, i.e. the ratio of the equivalent absorption surface area A to the volume of the room V (see "Room acoustics", p.15). According to VDI 2569, typical values are $A/V = 0.3$–0.35 m^{-1}. Reverberation times can also be derived and are then in the order of magnitude of 0.5–0.6 s. However, reverberation times are less meaningful in open spaces because the proportions of the room mean that there is no diffuse sound field.

The requirements regarding attenuation in the room relate to the octaves from 250 to 2000 Hz. The attenuation in the room can be somewhat less for lower frequencies, especially those below 100 Hz. At frequencies > 2000 Hz, the furnishings and fittings typically found in offices and the absorption effect of the air generally provide sufficient attenuation anyway.
The requirement placed on the A/V ratio (see above) is easily fulfilled when the sum of all the absorbent surface areas is approximately equal to the plan area of the room. This makes it clear that the form of construction that prevailed in the past, i.e. the use of suspended, sound-absorbing ceilings with the services routed

behind these, still has its advantages today: excessive reverberation and the disturbing propagation of sound via reflections off the ceiling are avoided. Ideally, the absorption of suspended ceiling systems should be as high as possible, e.g. $\alpha_w \geq 0.7$, better still $\alpha_w \geq 0.9$. Examples are mineral-fibre insulating panels in lay-in systems, perforated metal panels with a sound-absorbing backing, or perforated plasterboard. Cooling ceilings can also be designed to help absorb sound.

These days, exposed concrete soffits are by no means rare, which for the purpose of thermal activation of the suspended floor slab cannot generally be provided with an acoustic lining, or only to a limited extent only. Possible passive measures then include, for example, absorbent baffles or suspended ceiling panels combined with sound-absorbent ducts for the services (Fig. 2). Because however the amount of sound absorption required is generally not achieved with such methods, additional absorbent furnishings (movable partitions, cupboard fronts), possibly also wall linings, are necessary. Another advantage of such a soffit lining in the form of individual elements is that if the office structure is later changed to, for example, office cubicles, better sound insulation can be achieved with the walls because they extend up to the underside of the structural floor above, instead of terminating at a suspended ceiling.

The effect of carpeting on the floor is mostly overestimated because it only contributes to damping high-frequency sounds and – in contrast to an absorbent soffit – does not prevent sound propagation within the room. The only advantage of carpeting is with respect to the noise of footsteps within the room itself and also the fact that it reduces the transmission of impact sound to neighbouring rooms (p. 62).

It should be pointed out that an open-plan office can also be over-attenuated. This applies to rooms where fewer telephone and personal conversations are conducted simultaneously and where there is a low background noise level. High damping then leads to better speech intelligibility and hence makes it easier to overhear conversations conducted at both neighbouring and more distant workplaces.

Masking background noise
Whereas in the past mechanical ventilation systems, loud office equipment and also possibly windows that were not properly sealed resulted in a background noise level in offices of 40 – 45 dB(A), nowadays background noise levels of 30 dB(A) and even less are quite common. However, such "meaningless" noise helps to make the content of conversations at neighbouring workplaces less intelligible and hence less distracting. We speak of masking.
If a mechanical ventilation system is to be installed, it is not only worthwhile specifying a maximum noise level, but a minimum noise level as well!

It is possible to raise the background noise level marginally by introducing artificial sounds. Although this may sound simple, it is in fact difficult to generate a suitable background noise level that is consistent throughout the room. Playing music distracts just as much as conversations at neighbouring workplaces. It is also disturbing when the noise from loudspeakers can be localised or when the artificial noise level is too high; 40–45 dB(A) is about the limit here. Special sound-masking systems represent one solution. These have not become established in Germany's office landscape, but in the USA, for example, they have been in use for a long time. Retrofitting such systems in existing offices is not simple because

T1: Intelligibility of speech (loud speech level) from the next office depending on the weighted sound reduction index R'_w and the background noise level in the receiving room

Weighted sound reduction index of partition R'_w	Typical partition construction	Intelligibility of speech from next room with a background noise level of...		
		25 dB(A)	30 dB(A)	35 dB(A)
32 dB	8 mm lam. safety glass	very good	very good	good
37 dB	16 mm lam. safety glass with Si acoustic interlayer	very good	good	still intelligible
42 dB	100 mm plasterboard-clad stud wall	good	still intelligible	words intelligible but not whole sentences
47 dB	125 mm plasterboard-clad stud wall	still intelligible	words intelligible but not whole sentences	unintelligible, hardly audible
52 dB	155 mm plasterboard-clad stud wall	words intelligible but not whole sentences	unintelligible, hardly audible	inaudible

T2: Recommendations for normal and enhanced sound insulation in office buildings (within same lettable unit) according to DIN 4109 supplement 2 (Nov 1989)

Weighted sound reduction index $\geq R'_w$
Weighted normalised impact sound level $\leq L'_{n,w}$ (for floors, stairs)

Room function	Components	Minimum sound insulation		Enhanced sound insulation
General	Floors, stairs Floors to corridors and staircases	R'_w $L'_{n,w}$	≥ 52 dB ≤ 53 dB	≥ 55 dB ≤ 46 dB
Normal office activities	Corridor and office partitions Doors	R'_w R_w	≥ 37 dB ≥ 27 dB	≥ 42 dB ≥ 32 dB
Demanding mental activity or handling of confidential matters, e.g. between manager and secretary	Corridor and office partitions Doors	R'_w R_w	≥ 45 dB ≥ 37 dB	≥ 52 dB –

numerous loudspeakers – preferably concealed – have to be installed together with their cables. And if rooms are too reverberant and too loud owing to the lack of absorbent surfaces, the use of artificial masking to improve the situation is not sufficient on its own.

Screening
The third important acoustic element in open-plan offices is screening – in the form of movable partitions or furniture. The height of such elements should be at least 1.5 m and, in addition, the ceiling above the screening element must be absorbent in order to suppress acoustic reflections from the ceiling. The reduction

in noise level that can be achieved between neighbouring workplaces lies in the region of 3–10 dB.
If an unrestricted view is desired and there is already sufficient attenuation in the room, screens with glazed top sections are a possibility. Movable partitions or room-dividing systems also contribute to ensuring that the sound level of speech decreases as the distance increases, thus reducing the intelligibility of conversations at more distant workplaces.
If the ceiling above an open-plan office is completely reflective, full-height demountable room dividers are the only way of ensuring successful acoustics. Such room dividers are fixed at floor and ceiling

1   The best screening effect is achieved with full-height elements. If these elements include absorbent surfaces as well as glass, absorbent soffits may not be needed, depending on the interior layout.
2   Prefabricated partitions with a high proportion of glass can also achieve enhanced sound insulation to DIN 4109 supplement 2 for "normal office activities" provided they are carefully designed and constructed.

3   Alber Property Investment office, Lana, southern Tyrol, 2006, architect: Stefan Gamper
Reception zones also benefit from good acoustic attenuation. Slotted ceilings with an absorbent backing plus carpeting ensure the best conditions for communication. In the offices, partly perforated plasterboard ceilings plus absorbent cupboard fronts perform this function.

and can guarantee different levels of screening depending on their construction. If positioned to face the corridor, full-height screens ensure a degree of privacy and reduce distractions due to persons passing, slamming doors, etc.

Offices with enclosing walls

The undisputed advantage of offices within enclosing walls is the much better insulation against sounds from neighbouring areas compared to the open-plan structures described above. Which degree of sound insulation is necessary depends on the use of the room and the expectations of users. Tab. T1 helps to provide a subjective classification. The lower the level of noise in the room itself, the higher the level of sound insulation must be in order to guarantee the same subjective noise control standard. Although it is seldom possible to specify or influence the background noise level exactly during the planning process, a qualitative appraisal in advance should help when formulating the sound insulation requirements.

It is also possible to make use of, for example, DIN 4109 supplement 2, which contains recommendations for minimum and enhanced sound insulation, in both cases for "normal office activities", and for rooms where confidentiality is important (Tab. T2).

Airborne sound insulation for rooms without confidentiality requirements
A weighted sound reduction index of R'_w = 37–42 dB is generally adequate to ensure an undisturbed working atmosphere when no particular requirements are placed on confidentiality.

This level of sound insulation can be achieved with typical modern interiors, i.e. the use of walls in dry construction or prefabricated partitions built off hollow or raised access floors and/or suspended acoustic ceilings including adequate measures to protect against flanking transmissions. No complicated details are usually required for facades or the facade–internal wall junctions. Nevertheless, the architect is advised to specify the sound insulation values for all components (wall, floor, ceiling, facade, joints and junctions), which involve different trades, in order to achieve the overall noise control objectives.

As flanking components exert a considerable influence, it is not sufficient to demand a certain insulation value solely for the separating component.

Where interconnecting doors between offices are envisaged or are unavoidable because of fire requirements (alternative means of escape), the sound insulation of the wall is typically reduced to $R'_{w,res}$ = 32–37 dB (see "Building acoustics", p. 29). Cable ducts, leakage-air grilles and unsealed joints can also weaken the insulation significantly.

Where offices are located directly adjacent to open circulation zones or single-storey sheds, which have little acoustic attenuation and are hence louder, this may well lead to complaints.

Sound insulation for offices with confidentiality requirements
A weighted sound reduction index in the region of R'_w = 45–52 dB is necessary for offices and meeting rooms where confidentiality is important, i.e. nothing should be intelligible in an adjacent room. Such a level usually calls for totally different construction measures which have to be considered at an early stage of the planning. With enhanced sound insulation in particular (R'_w = 52 dB), the walls generally extend from the top of the structural floor below to the underside of the structural floor above in order to rule out flanking transmissions via floor finishes or suspended ceilings. Facades in post-and-rail construction also require special measures for this level of sound insulation ("Sound insulation between the units of different tenants", p. 64).

3

Somewhat higher requirements are placed on the sound insulation of doors in corridor walls (Tab. T2, p. 60). The same applies to doors between offices, e.g. between manager and secretary. If the best-possible sound insulation is to be achieved here, pairs of doors with R_w = 45 dB represent a good solution.

Reduced airborne sound insulation
A weighted sound reduction index of R'_w = 32 dB, which is less than the minimum requirement of R'_w = 37 dB given in the standards, can prove quite acceptable where mutual consideration is exercised and the background noise level is not too low. Such a standard is typical when the doors to the corridor of adjacent offices are left open – a situation common in everyday practice.

Reduced sound insulation standards are interesting because they can be realised with relatively simple constructions (e.g. 8 mm single glazing for partitions) and that results in corresponding cost-savings. Furthermore, a lower sound insulation standard is still much better than an acoustically optimised open-plan office. Such reduced sound insulation is also conceivable for the partitions between office cubicles in combined open-plan/cellular structures.

Impact sound insulation
In terms of the transmission of impact sound, a level of sound insulation 5 dB below the minimum sound insulation according to Tab. T2 (p. 60) is usually sufficient or everyday office situations, i.e. a weighted normalised impact sound level of $L'_{n,w}$ = 58 dB. The loadbearing structure can also be simplified in this case because stair flights can be connected rigidly to landings if necessary. A joint between office wall and stairs is, however, still necessary.

In the horizontal direction, i.e. from room to room or from corridor to room, carpeting on a solid floor slab, or even on a hollow floor, is usually sufficient. A hard floor covering (e.g. wood-block flooring, ceramic tiles), on the other hand, regularly leads to complaints concerning impact sound insulation. Even an isolating joint in a hollow floor seldom results in the desired effect because it can never be complete. As an alternative, or in addition, wood laminate flooring can be laid on a thin (5 mm) layer of impact sound insulation.

Room acoustics in offices
Besides using the right sound insulation to control noise between adjacent rooms, the room acoustics have an influence on users' well-being and the quality of their work even in smaller offices or meeting rooms.

In an office used by just one person, excessive reverberation leads to poor speech intelligibility at the other end of the line when telephoning.
A reverberation time of T_{target} ≤ 0.8 s can be regarded as a minimum acoustic standard, and this figure is often reached with the furnishings and fittings of a typical office. But with minimal furniture, paperless activities and no open shelves for files and books, the recommended reverberation time will certainly be exceeded, and the acoustic environment can give rise to complaints. The same is true for larger separate offices with a floor area > 15 m².

When the usage and the furnishings and fittings are not exactly definable during the planning phase, or alterations are possible, it is always advisable to provide a minimum amount of absorbent surface area – equal to about one-quarter to one-third of the floor area depending on the quality of the absorbent materials.

Using the guidance values given in DIN 18041, the additional equivalent absorption surface area introduced into an office used by just one person (area x sound absorption coefficient, see "Room acoustics", p. 15) should be at least 0.5 times the floor area. The figures are valid for a ceiling height of 2.5 m; the area of absorbent surfaces can be increased proportionately for taller rooms. When the recommendations of DIN 18041 are implemented, the reverberation time is T ≤ 0.6 s, even with minimal furnishings and fittings.

The sound-absorbent surfaces required are frequently provided in the form of a suspended ceiling over at least part of the office.
But where concrete slabs are designed to be thermally active, the use of suspended ceilings is restricted. The use of an absorbent lining over part of the soffit area or absorbent panels with a total area equal to about 30% of the floor space, possibly supplemented by absorbent materials on the walls, is common practice.
Sound-absorbent furniture can also be employed, but users must be made aware of the fact that it forms part of the room acoustics concept right from the very start.

1   Deutsche Kreditbank offices, Berlin, 2005,
     Kruppa Architekten
     Excessive reverberation in roofed-over inner court-
     yards, atria and circulation zones can be very dis-
     turbing, especially when special events are also
     held in those areas – a very common practice. In the
     atrium of the Deutsche Kreditbank offices in Berlin,
     absorbent timber cladding and acoustic plaster
     systems ensure the necessary acoustic attenuation.

In offices designed for between two and eight persons, adequate attenuation within the room is more important than in an office used by just one person because it is important to avoid a build-up in the loudness level during telephone conversations or direct personal communications. According to DIN 18041, the reverberation time in such offices should be about $T \leq 0.5$ s, although here again, individual usage scenarios can be taken into account when formulating the requirements.

Disturbing acoustic effects are unavoidable in such small multi-person offices when, for example, one person is telephoning and the others are trying to concentrate on their work. Panels mounted on the desks can act as screens, contribute to the attenuation in the room and also function as notice boards. The privacy of a separate office can, however, never be achieved.

Assuming that the workers in such an office are supposed to work as a team and communicate with each other, acoustically adequate working conditions can be achieved with sufficient attenuation within the room.

If the design is based on the requirements of DIN 18041, a lining to the soffit over the whole area is frequently necessary. However, it is also possible to transfer some of the absorbent surface area to the walls and furniture.

T3: Protection against sound propagation from work zones not within the same lettable unit; mandatory German building legislation requirements to DIN 4109 (Nov 1989) (excerpt)

Weighted sound reduction index $\geq R'_w$
Weighted normalised impact sound level $\leq L'_{n,w}$ (for floors, stairs only)

Room function	Situation	Requirements
Work zones not in same lettable unit	Walls between work zones not in same lettable unit	$R'_w \geq 53$ dB
	Floors between work zones not in same lettable unit	$R'_w \geq 52$ dB[1] $L'_{n,w} \leq 53$ dB
	Doors that lead from main corridors or stairs into corridors to work zones	$R_w \geq 27$ dB
Rooms with "especially loud" building services plant and installations; facilities for commercial operations, retail areas	Walls, floors enclosing rooms requiring protection	$R'_w \geq 57$ dB[2] or $R'_w \geq 62$ dB[3] $L'_{n,w} \leq 46$ dB[4]

[1] For buildings with more than two units: $R'_w \geq 54$ dB
[2] For max. sound pressure level in loud room: $L_A = 75-80$ dB(A)
[3] For max. sound pressure level in loud room: $L_A = 81-85$ dB(A)
[4] This value does not include machines on special bearings to attenuate structure-borne sound.

Good conditions for communications are imperative for meeting rooms. The most important design parameter is again the reverberation time. Taking DIN 18041 as our guide, a value of about $T \leq 0.9$ s (unoccupied) should be regarded as a maximum for a meeting room with a volume of 120–150 m³. However, in the case of more stringent demands, to take account of persons with impaired hearing and for using audiovisual equipment, the value should be somewhat lower, in the region of $T \approx 0.6$ s.

Measures similar to those used for offices can be used in meeting rooms as well. An absorbent lining to the rear wall, preferably covering the entire area, is also recommended. Larger meeting or conference rooms longer than about 10 m also require a reflective area in the centre of the ceiling to relay the soundwaves ("Lecture theatres, congress halls, plenary chambers", p. 78). Care must be taken to ensure adequate absorption of lower frequencies as well in order to achieve a balanced reverberation time. Otherwise, a seemingly muffled and booming impression ensues in rooms, especially those with carpet on the floor.

Sound insulation between the units of different tenants

The following information is relevant for office buildings containing units used by different tenants. Mandatory building legislation requirements, as given in DIN 4109, apply here, and are similar to those for multi-storey apartments (Tab. T3).

In new buildings, solid reinforced concrete floor slabs or walls around service cores are not critical from the acoustic viewpoint provided they are about 200 mm thick at least, which is frequently the case for structural reasons anyway. The walls to lift shafts should be 250 mm thick when these adjoin working areas directly.

The sound insulation requirements for walls between lettable units as called for by building legislation can be achieved, however, by plasterboard-clad double-stud walls or special acoustic boards and stud sections, provided these extend from the top of the structural floor below to the underside of the structural floor above.

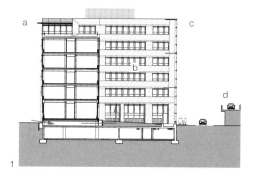

1

1, 2 Office building, Munich, 2001, DMP Architekten
A 22 m high delicate glass construction creates a
quiet inner courtyard directly adjacent to Donnersberger Bridge in Munich, one of the busiest inner-
city roads in Europe. The incident solar radiation
generates thermal currents in the courtyard which
draw in air from the rear of the building. The fresh-
air openings required are integrated into the base
of the building. Numerous offices can therefore be
ventilated naturally without being exposed to
noise and pollutants.
a Green zone, protected area
b Inner courtyard
c Glass screen, 22 m high
d Donnersberger Bridge, 160 000 vehicles/day

From the acoustics viewpoint, the facade usually requires special attention, particularly if it is a suspended post-and-rail construction. In the vertical direction, continuous posts and ineffectively clad rails represent potential weak spots. In the horizontal direction, the sound can be transmitted to the wall via the posts and the fin stiffeners. The glazing acts as a flanking element in both cases. If the mandatory sound insulation values of building legislation have to be complied with, this should be considered at an early stage of the facade design.

The impact sound insulation is usually less of a problem. For example, with a 200 mm deep reinforced concrete suspended floor slab, floor finishes with a weighted impact sound reduction index of $\Delta L_w \geq 20$ dB are adequate, a figure that is easily achieved by a hollow floor with resilient bearing pads. Carpeting can also reduce impact sound effectively. According to DIN 4109, carpeting may be included in the calculation provided there are no more than two units in the building.

Noise from outside
Office buildings are frequently located in central urban areas alongside busy roads. In order to limit effectively the interior noise level in offices caused by external noise, it is necessary to specify the sound insulation to be provided by the facade, generally by means of VDI 2719 "Sound insulation of windows and their auxiliary equipment".
With high external noise levels, continuous natural ventilation via partly open windows is frequently impossible owing to the high sound immissions. This problem can be dealt with by employing a suitable facade design and interior layout.

Double-leaf facades
Double-leaf facades designed to meet acoustic requirements permit natural ventilation even on busier roads. The second, outer facade leaf increases the sound insulation and can ensure an acceptable noise level in the interior so that communication is possible. Flanking transmissions between neighbouring rooms (above, below or to the sides) can be reduced to a tolerable level by constructional measures, e.g. horizontal and vertical baffles.

"Hybrid windows"
An example of a building using hybrid windows is shown in the chapter on "Noise control in urban planning" (p. 46, Figs. 1 and 2). This type of window exploits the advantages of the double-leaf facade without suffering from its disadvantages.

Glazed screens
Very effective from the acoustic point of view is the use of a tall glass screen to create a quiet inner courtyard. Depending on their construction and height, such glass screens can achieve a reduction in sound level of 15–20 dB.

2

Swiss Re offices, Munich

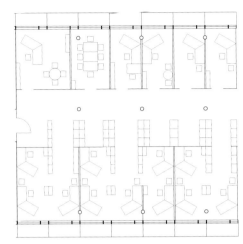

Architects: BRT Architekten,
             Bothe Richter Teherani, Hamburg
Acoustics
consultant: Müller-BBM, Planegg
Completed: 2002

This new development consists of two separate buildings with different geometries: a two-storey plinth containing shared facilities such as conference zones, library, employees' casino and visitor area; raised above this on stilts are smaller structures containing a total of 16 office units on two levels.

The concept of a central service core and office units branching off from this forms the foundation for creating good acoustic conditions in the working areas. This arrangement allows meeting rooms, cafeterias and sanitary facilities to be located centrally but nevertheless directly adjacent to the work zones. Annoying distractions caused by noises from the cafeterias, slamming doors and other disturbing activities are therefore ruled out. Providing access to the office units from one side only means that "through traffic" is eliminated.

Furthermore, the layout of the units ensures that all the workplaces benefit from the same qualitative features. Every workplace commands a view of the gardens outside and the planting on the flat roofs, the hedge surrounding the building or the internal courtyard.

Open-plan office landscapes are broken up alongside the facades by full-height demountable partitions which include absorbent surfaces. The sound-absorbent ducts for services in the longitudinal and transverse directions also house indirect lighting. Thanks to a spacious internal layout, additional absorbent surfaces on the soffits proved to be unnecessary. Instead, the additional internal attenuation required is provided by furniture with absorbent front panels.

Offices for one or more persons are formed by prefabricated partitions with excellent acoustic properties and large proportions of glass.

Part-plan of office storey   scale 1:300
Section • Plans   scale 1:1500

1   Entrance, foyer
2   Auditorium
3   Library
4   Fitness centre, clubroom
5   Casino
6   Kitchen
7   Restaurant for guests
8   Cafeteria
9   Photocopiers
10   WCs
11   Meeting room
12   Stairs, lifts

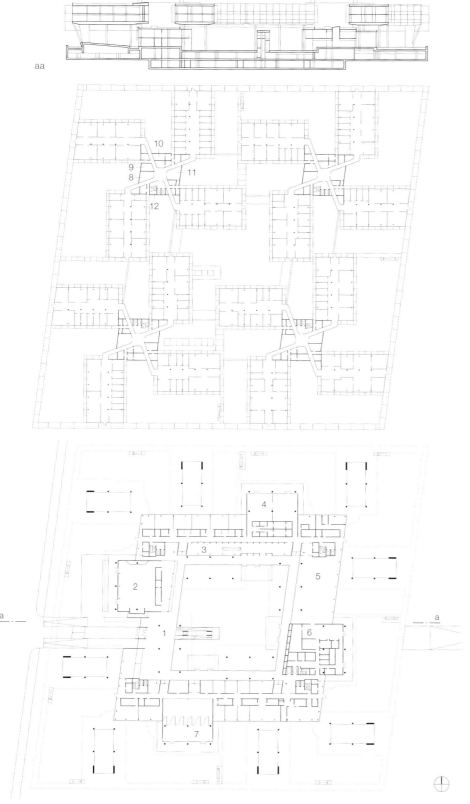

Learning and teaching conditions in schools and other educational establishments are considerably influenced by the construction of those buildings. Light, air quality, interior climate and acoustics are factors considered in the course of refurbishment work, extensions and new buildings.

This chapter will explore the need for good acoustic conditions, the underlying statutory instruments, standards and directives, and constructional solutions. However, the interaction between, and partly conflicting requirements of, acoustics on the one side and air quality and interior climate on the other will also be investigated (Fig. 1). What use is, for example, optimum thermal comfort when in the classroom it is too reverberant and loud because of numerous fair-face concrete surfaces (thermal mass)? And what use are the best sound-insulating windows when they have to be opened despite the high external noise level because there are no alternative ventilation options?

This chapter focuses on schools for primary and secondary education. Many of the relationships described, however, can be transferred to other educational establishments such as training colleges and adult education centres. On the other hand, in nurseries, kindergartens and similar child-care facilities, noise development and exposure are important.

Room acoustics in schools

About 75% of the time spent in classrooms is devoted to speaking and listening. Obviously, the constructional conditions should therefore take spoken communications into account. But the reality is often very different. In many older school buildings, but even in newer ones, acoustic design plays little or no role at all, which results in excessive reverberation in classrooms. Such conditions minimise,

subjectively and objectively, the intelligibility of speech (Fig. 3), and – perhaps even more important these days – reverberant rooms encourage restlessness among the pupils. Studies and empirical evidence show that in rooms without any acoustic treatment, the noise level induced by the users is up to 10 dB higher than is the case in rooms with acoustic treatment. The intelligibility of speech, particularly in the back rows, is therefore reduced, forcing the teacher to speak louder, which in turn increases the stress on vocal chords and psyche. It is not without reason that medical insurers are becoming increasingly interested in the subject of acoustics in classrooms. Children with impaired hearing or those being taught in a language that is not their native tongue suffer especially in classrooms with excessive reverberation.

Such findings are in no way new, but they have been underpinned by numerous studies in recent years, and the public is becoming more aware of them in the wake of the "Pisa Study" and the German government's "education offensive".

Desirable reverberation times
The reverberation time is the most important criterion for the quality of the room acoustics in a classroom. Recommended values are given in DIN 18041 "Acoustic quality in small to medium-sized rooms", which in the meantime can be regarded as a "generally acknowledged technical standard", especially for the design of classrooms.

According to this standard, the reverberation time in an occupied classroom for the room volumes typically encountered – approx. 150–250 m³ – should lie in the range T_{target} = 0.5–0.7 s. These values are especially relevant for the octaves from 250 to 2000 Hz – the range in which the critical energy components of human

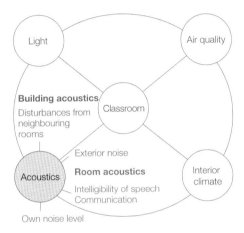

1a

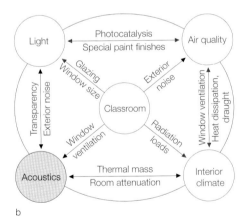

b

1 a Acoustic aspects in school buildings
  b Interaction between interior climate, air quality and light
2 Sound-absorbent ceilings in classrooms
  a New secondary school in Grafenau, 2004, Klaus Bauer architectural practice and Heinrich Scholz architectural practice: mineral boards with perforated metal facing
  b Laubegg School, Winterthur-Dättnau, 2002, Roland Meier, with Marc Schneider & Daniel Gmür: wood-wool acoustic boards
3 Relationship between reverberation time T and intelligibility of speech in classrooms
  a T = 0.5 s
  b T = 1.0 s
  c T = 1.5 s
  With a reverberation time of 1 s, objectively good intelligibility of speech is no longer guaranteed; values of 0.5 – 0.7 s are regarded as an optimum.

2a   b

speech are to be found. Whether the upper or the lower value is used as a reference point during the planning depends on the situation. In primary schools and also classrooms for persons with impaired hearing, the reverberation times should tend towards the bottom end of the afore-mentioned range.

Starting with the desired reverberation time and the sound absorption of the pupils' clothing, it is possible to estimate the quantity of absorption surface area required based on the statistical reverberation theory (see "Room acoustics", p. 15). Typical values lie in the range 50–80 m^2, which roughly corresponds to the plan area of the room.

This information supplies a ballpark figure and can vary up or down depending on the materials and form of construction used. The areas can be defined more accurately by way of frequency-related reverberation time calculations, also taking into account the planned or existing furnishings and fittings, in order to enable, for example, better coordination with architectural, interior climate and economic aspects. During refurbishment work, measuring the reverberation times in the existing building is a useful expedient that provides sensible, accurate information for planning.

Distribution of absorbent surfaces
The most popular solution is to provide classrooms with sound-absorbent ceiling systems. Fleece-backed and colour-coated mineral ceiling panels, in a lay-in suspended ceiling system or attached directly to framing below the soffit, are very widespread, along with perforated plasterboard with an appropriate backing. Fig. 2 above and Fig. 1 on p. 70 show a few examples of ceilings already installed. If the acoustic treatment is restricted to the ceiling, it is important to ensure an

adequate suspension height so that lower frequencies are also absorbed.

The significance of "low frequencies" has been hotly debated for some time. New findings allow us to conclude, however, that a rise in the reverberation time of up to 20% in the 125 Hz octave is quite acceptable. Another reason for the interest shown in this is the fact that the absorption of low frequencies frequently calls for additional constructional input, which means extra costs.
A sound-absorbent lining over the entire soffit as the sole acoustic measure is not the first choice from the acoustics viewpoint. In the majority of existing rectangular plan forms, multiple reflections build up between the parallel walls, especially with minimal furniture; such reflections impair the intelligibility of speech and in extreme cases can be perceived as disturbing flutter echoes. At the very least, however, concentrating the absorbent surfaces on the ceiling frequently leads to the reverberation times being longer than predicted (see p. 15).
These disadvantages can be dealt with by providing some absorbent surfaces on the rear wall, e.g. an absorbent notice board (see "Room acoustics", p. 23).

In new buildings, a lining covering the entire soffit usually conflicts with the interior climate requirements. If a large part of a solid floor slab acts as a thermal mass, in conjunction with night-time cooling it can help to curtail summertime temperature peaks. From the acoustics viewpoint, we then require a considerable area of absorbent material on the rear and side walls amounting to about 30% of the plan area of the room in addition to sound-absorbent surfaces on the ceiling. A detailed case study (primary school in Erding) can be found at the end of this chapter (pp. 75–77).

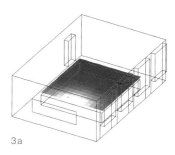

3a

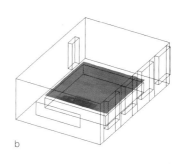

b

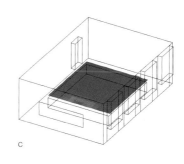

c

Speech Transmission Index (STI)

0,8 — very good
0,7 — good
0,6 —
0,5 — satisfactory
0,4 — poor

1a   b

Individual suspended ceiling panels (not part of a suspended ceiling system), as are common in offices, are rare in schools because of hygiene considerations (dust).

Specialist classrooms
The aforementioned relationships can be readily transferred to specialist classrooms. Furthermore, in large classrooms or specialist rooms more than about 10 m long, it is desirable for early reflections to be reflected off the ceiling to the back rows. This means that the middle part of the ceiling must be designed to be reflective (see "Lecture theatres, congress halls, plenary chambers", p. 78, Fig. 1). Contrasting with this, in noisy workshops or similar facilities, adequate acoustic attenuation in the room itself is the priority, which can be achieved very well with an absorbent ceiling system over the entire area.

Rooms for teaching music must be approached differently. High attenuation in the room is desirable for drums, percussion and electroacoustically amplified instruments, whereas somewhat longer reverberation times tend to be favoured for string, woodwind instruments, and pianos.
Here again, individual tastes and practices play a major role, and the main functions of the room should be taken into account, particularly in the case of refurbishment and conversion measures; variable acoustic measures may be necessary (see "Small rooms for music", p. 82).

Gymnasiums
In a gymnasium, sound-absorbent measures are used to reduce the reverberation and the noise level generated by users themselves. The reverberation times

2

3a   b

desirable for gymnasiums depend on the volume of the room and are given in DIN 18041, to which the current edition of the "gymnasium standard" (DIN 18032-1 "Halls and rooms for sports and multi-purpose use – Part 1: Planning principles") refers. Maximum reverberation times are given as 1.8 s for a single gymnasium and 2.5 s for a divisible gymnasium.

Large areas of absorbent materials are required to match the large volume of a gymnasium, and each case has to be calculated individually. These areas are frequently provided in the form of slit, slotted or perforated wood-based board products or plasterboard, wood-wool acoustic boards, mineral insulating boards, etc. on the ceiling, which are offered by the manufacturers in versions capable of resisting ball impacts. But limiting the sound-absorbing surfaces to the ceiling alone is never sufficient; the height of the gymnasium, the rectangular plan form and the acoustically smooth wall surfaces result in significantly longer reverberation times and also disturbing flutter echoes when absorbent materials are used nowhere other than the ceiling. It is necessary to provide absorbent surfaces on the rebound

zones of walls as well. To reduce the risk of injury, hole diameters or the gaps between spaced components may not be larger than 8 mm.

In a divisible gymnasium, absorbent, double-layer curtains to divide up the area can help to improve the room acoustics. Gymnasiums are frequently also used as school assembly halls or for other events. It is therefore very often the case that a stage is provided at one end of the room, e.g. separated by a movable room divider. The acoustic design must then take the multi-functional character of the gymnasium into account as well. For example, besides a suitable reverberation time, the transmission of reflections from the stage to the audience is also relevant.

Lunchtime facilities
Germany's education reform measures include the provision of dining areas and other lunchtime break amenities – in the form of new extensions or the conversion of existing rooms. The primary task of room acoustics in this case is to limit the noise level, caused by conversations, the use of crockery and cutlery, etc., which calls for sufficient absorptive surfaces to ensure short reverberation times. In particular, a ceiling with good absorption properties is necessary, the detailed design of which is determined depending on the architectural and functional requirements.

The task becomes more difficult when the rooms also have to be used for school events in addition to lunchtime breaks. The demands for high attenuation within the room and a reduction in the sound level conflict with the demands for a good natural acoustic, i.e. good sound propagation and adequate resonance. This contradiction can succeed when – as with the use of a gymnasium for multiple functions – the room is sufficiently high so that

a minimum volume of about 7 m³ per person is available for events. A certain degree of resonance can then be achieved for school events and the bigger room means that the noise of users during lunchtime breaks is distributed over a larger volume.

Circulation zones
Acoustic damping in corridors serving classrooms helps considerably to ensure that the build-up of noise as pupils move around the school remains within acceptable limits. A high noise level in a corridor is particularly disturbing for a class that is still being taught in an adjacent classroom. For this reason, architects would be well advised to provide absorbent surfaces in corridors, e.g. in the form of suspended ceilings (Figs. 3a,b).
The same applies to indoor areas used at break times, and in such cases the requirements of multi-functional usage may need to be taken into account as well (p. 73, Fig. 1).

Nurseries and similar child-care facilities
In preschool facilities, a reverberant environment can sometimes lead to an escalation in restlessness among the children to an even greater extent than in (primary) schools. Although plants, carpets, furniture and toys do contribute to attenuating the noise in the room, they are not sufficient on their own. Sound-absorbing surfaces can achieve decisive improvements. The objective planning variable here is again the reverberation time, which should, ideally, be somewhat lower than that in primary schools, i.e. $T_{target} = 0.4 - 0.5$ s (room occupied).
Absorbent ceiling systems with a sufficiently high absorptive effect and, wherever possible, over the full area of the room are ideal for internal attenuation in preschool facilities.

1 a, b Refurbishment of a secondary school in Aichach, 2002, Obel + Partner
   Classroom after refurbishment with ceiling of perforated plasterboard (8 mm dia. holes, 18 mm pitch), with absorbent backing. The curving ceiling and the change in level along the sides create additional absorption surface area.
2   Gymnasium/multi-purpose hall at the European School in Munich, 2004, Moosmang + Partner
   The stage is normally concealed behind a movable room divider. The curving ceiling guarantees a good distribution of reflections for events. Panels along the sides of the ceiling conceal sound-absorbent surfaces behind beam grids. The fluted wooden lining creates a uniform image, but it is in fact partly reflective, partly absorbent (full-depth slits).
3   Absorbent linings in corridors help to combat disturbing, high noise levels.
   a Perforated plasterboard
   b Metal grid

T1: Sound insulation requirements in "schools and comparable educational establishments" according to DIN 4109 (Nov 1989

Component	weighted sound reduction index reqd. R'_w (for doors: R_w)	weighted normalised impact sound level reqd. $L'_{n,w}$
Floors between classrooms or similar rooms	≥ 55 dB	≤ 53 dB
Floors below corridors		≤ 53 dB
Floors between classrooms or similar rooms and "especially loud" rooms (e.g. gyms, music rooms, workshops)	≥ 55 dB	≤ 46 dB
Walls between classrooms or similar rooms	≥ 47 dB	
Walls between classrooms or similar rooms and corridors	≥ 47 dB	
Walls between classrooms or similar rooms and "especially loud" rooms (e.g. gyms, music rooms, workshops)	≥ 55 dB	
Doors between classrooms or similar rooms and corridors	≥ 32 dB	

Rooms for physical activities should also be provided with suitable absorbent surfaces to reduce the noise level.

Such acoustic aspects are easily included in the course of renovation or refurbishment work.

Sound insulation within the building

Mandatory building legislation requirements exist for "schools and comparable educational establishments", and these are specified in DIN 4109 "Sound insulation in buildings". The objective of the provisions is to prevent "unreasonable disturbances" caused by sound immissions from neighbouring rooms, from outside and from building services.

Frequent acoustics problems in schools due to the building acoustics are as follows:

- Disturbances caused by footsteps and the scraping of chairs owing to poor impact sound insulation (especially in older buildings)
- Disturbances caused by "loud classrooms", especially music rooms, owing to inadequate airborne and impact sound insulation
- Poor airborne sound insulation between

classrooms with interconnecting doors (alternative means of escape), continuous ventilation or service ducts passing through the rooms, poor workmanship at junction details

- Disturbances in classrooms due to loud noises in adjacent corridors
- Disturbances caused by excessive external noise on busy roads, or conflicts with natural ventilation

These and other acoustic deficits due to the design and construction of the building itself can usually be avoided if the mandatory building legislation requirements are taken into account and implemented. The requirements are listed in Tab. T1.

A number of typical constructions suitable for separating components in schools and similar educational establishments are described below; these comply with the building legislation requirements in principle. Possible solutions to overcome problems in existing buildings are also discussed. The information is primarily intended to act as a general guide and should not be understood as "patent recipes". In each specific application, an

acoustic evaluation by a competent person, taking into account all influencing factors, will lead to a reliable design.

Walls between classrooms
The sound insulation necessary for walls between classrooms can be provided by, for instance, the following forms of construction:

- ≥ 110 mm reinforced concrete
- ≥ 175 mm masonry, density class 1.4, plastered both sides
- 150 mm wall in dry construction, characteristic value for wall without flanking transmissions $R_{w,R} \geq 52$ dB

It is generally necessary to build the walls off the structural floor below and continue them to the underside of the structural floor above.

Another condition is that flanking components (ceiling, floor, facade, corridor wall) should not constitute acoustics weaknesses. This is especially the case when solid components exhibit an average mass per unit area of 300 kg/m². Double-leaf (non-rigid) lightweight walls are not usually critical with respect to flanking transmissions.

On the other hand, curtain-wall facades with a post-and-rail construction always require more accurate acoustic design and detailing to achieve the level of sound insulation necessary in a school building. Potential weaknesses are again service or ventilation ducts where these pass directly from one classroom to another. However, this problem can also be overcome when the acoustic aspects are given due attention during planning and tendering procedures.

If a door is required between two classrooms to provide an alternative means of escape, this always represents a weakness and so should be avoided wherever possible. Should that prove unfeasible,

the building legislation requirements cannot be complied with in most cases and even the installation of high-quality sound-insulating doors with a weighted sound reduction index of $R_w \geq 37$ dB cannot prevent a drop in the quality of the sound insulation between the rooms. The situation can be solved fairly simply if a narrow storage room, e.g. for storing teaching materials, can be included between the two classrooms, as is often the case between specialist science classrooms.

The sound insulation to classrooms for music in existing schools is very often critical where these adjoin other classrooms. Whether in such a case a non-rigid independent wall lining built in front of the separating wall proves helpful depends quite crucially on the flanking components. It is not possible to specify a universal answer.

Walls to corridors
According to DIN 4109, walls between classrooms and corridors are subject to the same airborne sound insulation requirements as walls between classrooms, i.e. the same or similar forms of construction can be employed.
But what options are available when there are fanlights over the doors or high-level glazing in the wall? (This is practically always the case in new buildings or extensions.)
When the fanlights above the doors are small, the sound insulation can be attributed to the door elements, i.e. a weighted sound reduction index of $R_w \geq 32$ dB (Tab. T1) is required for the door plus glazed element (or laboratory value: $R_{w,P} \geq 37$ dB). Single glazing approx. 8 mm thick is normally adequate.
The situation is more difficult with a large area of high-level glazing. Here, depending on the wall construction, elaborate, costly sound-insulating glazing units, even two separate planes of glazing (coupled

windows), may be necessary in order that the wall as a whole – consisting of opaque and transparent parts – achieves the weighted sound reduction index of $R'_w \geq$ 47 dB required by building legislation. The question often arises as to whether single glazing can be used instead. The answer from the building legislation viewpoint is usually no. But if other aspects are taken into account, it can be argued that including absorbent materials in the corridor reduces the ensuing noise level anyway, which in turn justifies the use of sound insulation with lower values. Alternatively, the doors can be upgraded acoustically so that the total sound insulation satisfies the building legislation requirements. Such solutions must be assessed in every individual case before a plausible constructional solution can be proposed to the client for him to make a decision.

Suspended floors and floor finishes
The requirements regarding the airborne and impact sound insulation of suspended floors for classrooms (p. 72, Tab. T1) are fulfilled by solid structural floors with a mass of about 300 kg/m² in conjunction with a floating screed (e.g. reinforced concrete 130 mm deep with a floating screed $\Delta L_w \geq 26$ dB, dynamic stiffness of insulating material SD ≤ 20 MN/m³, floor finishes at least 80 mm deep).
With structural floors having a lower mass per unit area, e.g. older school buildings with ribbed floor slabs and also those with timber joist floors, the sound insulation usually needs improving. During refurbishment work it is advisable to establish the

1   Secondary school in Traunreut, 2003,
    Rainer A. Köhler architectural practice
    Area for indoor breaks and events:
    Conflicting acoustic requirements can be harmonised by using absorbent wall surfaces and ceiling panels, and by ensuring an adequate room volume.

situation by way of measurements or by referring to the original drawings. Armed with this information, an analysis can be carried out to establish whether and which acoustic upgrading measures are necessary. Possible solutions include floating screeds and ceilings comprising two or more layers of plain plasterboard on resilient bars (an initial measure to which a sound-absorbing suspended ceiling can be added).

A problem often encountered with floating screeds in old buildings is the minimal depth available for floor finishes. But with ceilings, fire requirements can also be satisfied. In older buildings in particular, with intermediate floors exhibiting poor acoustics but at the same time thick (and heavy) masonry walls, this represents one option.

Timber construction
In principle, acoustic requirements for school buildings can also be met by timber buildings, although this does require some effort and besides floating screeds also calls for independent wall linings in front of the main walls.

Sound insulation for "especially loud rooms"
DIN 4109 specifies higher requirements for the sound insulation between especially loud rooms and other rooms requiring protection from noise. The especially loud rooms in schools include workshops, gymnasiums and music rooms.
To satisfy the $R'_w \geq 55$ dB requirement, wall constructions comprising, for example, 220 mm reinforced concrete, or 240 mm masonry, density class 2.0, plastered both sides, or double-stud walls in dry construction are necessary.

Solid structural floors should have a mass per unit area of about 450 kg/m² (e.g. 200 mm reinforced concrete). A floating

screed is also necessary ($\Delta L_w \geq 26$ dB, min. 80 mm depth of floor finishes). Experience has shown that the building legislation requirements for music rooms – depending on usage – are comparatively low. In new construction projects, but also during refurbishment work, besides ensuring a suitable layout of the rooms within the building, a check should also be carried out as to whether and to what extent a higher standard of sound insulation can be achieved, which could be based on the standard used, for example, at music colleges (see "Small rooms for music", p. 82).

Insulation against external noise
Depending on the level of external noise, for an educational establishment proof of protection against sound immissions will have to be submitted to the building authority for approval. The facade and the windows will have to be acoustically designed to suit the critical external noise levels in front of the facade. This means, for example, that windows of a certain sound insulation class must be installed. However, the question that then always arises is: How should the classrooms be ventilated?
It has been shown that ventilating the rooms via doors and windows only during the breaks between lessons does not ensure an adequate air quality in the classrooms.
One potential solution, excellent from both the technical and economic viewpoints, is to assist the inflow of fresh air by providing sound-insulated fresh-air inlets in the facade (see case study, p. 77). Such concepts can also be pursued in refurbishment projects for existing buildings and facades, and supply good results for external noise levels up to about 70 dB.

During the design work, make sure that

the fresh-air inlets do not impair the sound insulation of the facade significantly, i.e. that the inlets specified in the tender have an adequate normalised element sound level difference. The value required depends on the external noise exposure, the number of inlets and the sound insulation of the windows and facade.
Low-noise wall-mounted ventilation fans are being increasingly considered as more and more school facades are refurbished. These do not achieve the air change rates of about 20 m³/h per person necessary for good air quality because the noise level of the fans themselves would then be too high. However, in existing buildings the lower air change rates in conjunction with opening the windows between lessons achieves a noticeable qualitative improvement.

Primary school in Erding

Architects: Dinkel Persch, Erding
              Wollmann & Mang, Munich
Acoustics
consultants: Müller-BBM, Planegg
Completed: 2005

A

Schools, or rather education establishments in general, usually place great demands not only on acoustics, but on the whole spectrum of building physics. More than with any other type of building, the most diverse, and in some instances conflicting, requirements have to be fulfilled. Besides the acoustic issues (good room acoustics, high sound insulation for the rooms where larger groups of people come together to learn), both the thermal and visual comfort and, in particular, the quality of the air are extremely important. Numerous studies, including those carried out recently, indicate the massive deficits in the existing school building stock.

One good example of how an holistic, or rather integral, approach to planning can satisfy the most diverse requirements is the new primary school in Erding.

As the school is situated directly adjacent a busy road between Erding and Altenerding, special noise control measures were necessary. Two buildings – the school building itself and the gymnasium – screen the schoolyard against external noise and at the same time protect the nearby housing against sound emissions from the school's outdoor areas.

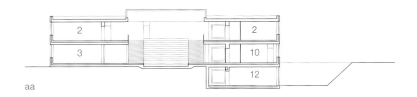
aa

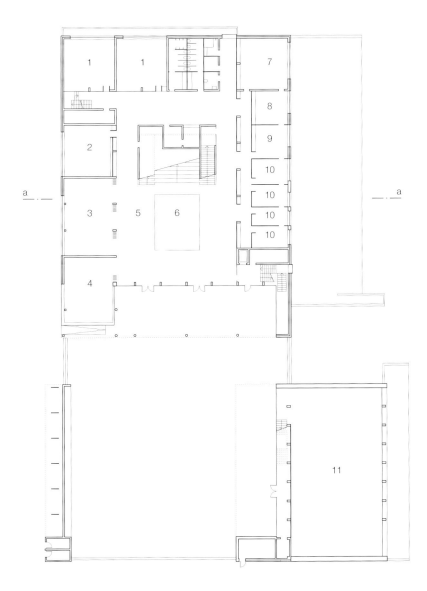

A   Facade facing busy road:
    Despite generous areas of glazing, the noise of
    the road does not infiltrate the school building
    thanks to sound-insulating windows and sound-
    insulated fresh-air inlets.

Section		5	Auditorium
Plan of ground floor		6	Seating area
scale 1:600		7	Staff room
		8	Homework/tutorial room
1	Lunchtime break room	9	Library
2	Classroom	10	Office
3	Multi-purpose room	11	Gymnasium
4	Music room	12	Workshop

B

Measures were also necessary on the building envelope itself. Sound-insulating windows were installed in the facades facing the road. The ventilation concept, too, had to take account of the external noise. With such framework conditions, ventilation via the windows could not even be considered. In order to minimise the technical complexity, sound-insulated fresh-air inlets in the facade plus central mechanical exhaust-air extraction were installed instead of a conventional ventilation system. The much lower flow resistance results in energy-savings for the transport of the air as well.

Fresh external air is therefore introduced via four inlets per classroom. The stale air in the room is extracted via sound-insulated outlets in the corridor walls – which lead to a main duct in the hall used at break times – and is vented to the outside via the roof in the central atrium.

The inlets were specially developed for this project and their acoustic aspects optimised in such a way that the low-frequency noise of passing HGVs is ade-quately attenuated. The noise attenuation of the inlets are assisted by the facade of ceramic mosaic tiles with ventilation cav-ity, which is designed as a "silencer". The thermal insulation also helps to absorb the sound. Another advantage of this ven-tilation concept is to be found in the high quality of the interior air. Measurements in schools have revealed that an adequate air quality cannot be guaranteed with ventilation via the windows because suffi-ciently intensive continuous ventilation is impractical, especially in winter. By contrast, the inlets at the primary school in Erding are designed in such a way that there is no serious loss in comfort even when the temperature outside is low. Furthermore, the ventilation system plays a key role in the interior climate concept. In summer the ventilation also operates during the night so that cooler external air can flow through the classrooms and dis-sipate the heat stored in the solid compo-nents of the building during the day. To achieve a pleasant interior climate in the summer, the night purging of the building's thermal masses is crucial. This fact is often overlooked, even with new buildings.

In order to avoid excessive solar gains in the summer, the window sizes are limited to that required to provide good daylight conditions in the interior; they are also fitted with louvre blinds to provide shade from the sun.

Solid components have been provided as thermal masses to curtail the peak summer-time temperatures in the interior. This is where the climate concept affects the acoustic concept. In order that the thermal masses can be exploited, most of the fair-face soffits to the concrete floor slabs have been left exposed. The sound-absorbent surfaces necessary to attenu-ate the sound within each room have been installed as a peripheral ceiling panel and placed on the rear wall and the wall adjacent the corridor. The wall linings can also be used as notice boards for class and school activities. More recent studies have shown that this distribution of the sound-absorbing surfaces – com-pared to a conventional acoustic ceiling – improves the intelligibility of speech and does not result in any disadvantages for the acoustic environment, indeed can even be advantageous. Disturbing reflec-tions due to the parallel walls are there-fore suppressed from the outset and the exposed fair-face concrete soffit helps to ensure a cooler interior climate in the summer.

As well as serving as a circulation zone, the central atrium, around which the classrooms are arranged, provides an indoor recreational area during breaks and a venue for school events. To reduce the noise level, the soffits in some areas were clad with a sound-absorbent mate-rial. Another benefit of this is that it cre-ates favourable acoustic conditions for many types of event.

The fact that the building has proved its worth in everyday use and is very popular

C

D

Section through fresh-air inlet on facade
scale 1:10

1 Opening light,
   aluminium window with sound-insulating glazing
2 Wall construction:
   coloured ceramic mosaic tiles, 2 colours combined,
   25 x 50 mm, bonded to facade panel,
   24 mm recycled glass granulate
   80 mm ventilation cavity
   100 mm mineral wool thermal insulation with
   fleece facing

   250 mm reinforced concrete
   5 mm skim coat finish
3 Fresh-air inlet with internal sound baffles
4 Manually operated ventilation grille
5 Floor construction:
   2.5 mm linoleum, welded seams
   65 mm cement screed, separating layer
   20 mm impact sound insulation
   60 mm insulation as levelling layer,
   PE sheeting, 280 mm reinforced concrete
6 Ceiling panel: 25 mm wood-wool acoustic boards,
   50 mm mineral wall backing

with pupils and teachers alike is the result
of intensive communications between
client, architects and the individual spe-
cialists, right from an early stage of the
planning. Only when all those involved in
a construction project work together and
remain open-minded in the interdisciplinary
discussions can such a happy symbiosis
of architecture, building physics and
building services be created.
In the light of the ever-increasing pressure
to shorten the planning phase plus massive cost-savings in the planning process,
this should make us think again. Holistic
solutions are rewarded by a high degree
of satisfaction among users and frequently
by lower investment and operating costs
as well. However, this normally requires a
somewhat higher input at the planning
stage. The primary school in Erding proves
that the effort is worthwhile.

B The school building and the gymnasium screen
   the schoolyard against external noise and at the
   same time protect the neighbouring housing
   against sound emissions from the school's outdoor areas.
C The room acoustics measures in the classrooms
   are hardly noticeable and consist of a peripheral,
   absorbent ceiling panel made from wood-wool
   acoustic boards. This solution allows the thermal
   mass of the floor slab to remain effective. The
   sound-insulated fresh-air inlets are located
   between the radiators.
D Gymnasiums require absorbent surfaces on the
   ceiling and wall surfaces to reduce reverberation
   and noise levels. Reinforced concrete beams,
   radiant heating panels and wood-wool acoustic
   boards alternate at ceiling level. Timber battens
   spaced apart on rebound walls, in conjunction
   with an absorbent backing, prevent disturbing
   flutter echoes.

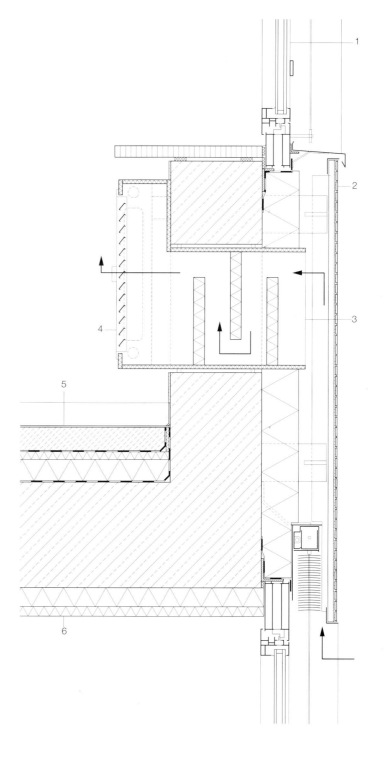

Lecture theatres, congress halls, plenary chambers

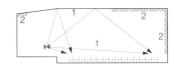 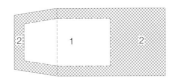

1a   b

The interior spaces studied in this chapter have two things in common: good speech intelligibility and good visual contact with the speaker or a screen are crucial for the function of the room.

Shape of room and distribution of absorbent surfaces
Many different forms are used for this group of rooms: rectangular, fan-shaped or arena-style plan layouts are just some examples. The rows of seats are positioned to face the podium, and laid out on a rake – steep or shallow – to suit the size of the room. Such an arrangement achieves not only good sightlines, but also good direct sound transmission to each person in the audience.

To improve clarity it is also very important that the geometric and acoustic design of the ceiling is such that early reflections reach the audience. This means that the front, central part of the ceiling should be designed to reflect the frequency range significant for speech (octaves from 250 to 2000 Hz). The borders and rear part of the ceiling can include absorbent materials to assist the attenuation within the room (Fig. 1). Ideally, the rear wall should also be sound-absorbent, at least over the area starting 1 m above the floor and especially when the seating is not raked. An absorbent rear wall will prevent delayed reflections, which hinder speech intelligibility, from being reflected back to the podium.

Extent of absorbent surfaces
Besides assistance for early reflections, good speech intelligibility also needs a low reverberation time, which with a room volume of, for example, 1000 m³ should be about 0.8–1 s according to DIN 18041. An excessive room volume per person is a hindrance when trying to achieve such a reverberation time; a figure of about

4–6 m³ per person is regarded as ideal. The audience and/or the upholstered seats can then account for a large proportion of the sound-absorbing surface area required, and it is usually sufficient to allocate about two-thirds of the plan area of the room to sound-absorbing measures. For example, in the library and lecture theatre building in Weimar, reflective and absorbent surfaces are concealed behind timber battens on the walls and ceiling in the 400-seat lecture theatre with a volume of about 2500 m³ (Fig. 2).
It is important to provide sufficient damping for the 250 Hz octave in particular and not only for the high and medium pitches. Conventional carpeting on its own is not good enough for this.

Electroacoustic amplification
It is very difficult to provide a general answer to the question of: "at which room volume does electroacoustic amplification become indispensable?" Experienced speakers can make themselves heard in room volumes of up to about 2500 m³ without amplification, provided the background noise level is sufficiently low. But these days even small lecture theatres and conference rooms are equipped with sound reinforcement systems, partly due to the increasing use of multimedia devices (video projectors, audio equipment).

With a sound reinforcement system, there is no technical limit to the size of the room from the acoustic viewpoint. However, the presence of electroacoustic amplification in no way means that the natural acoustics are no longer relevant and sound-absorbent surfaces can be omitted. Quite the opposite in fact: the requirements regarding the damping in the room are even higher because it is important to counteract feedback from and fluctuations in the sound reinforcement system in order to achieve good-quality sound

(Fig. 3). Furthermore, care must be taken to ensure that reflections from any remaining reflective surfaces, e.g. a window to a control room located in the rear wall, do not become disturbing echoes.

In the National Convention Centre in Hanoi, for instance, the enormous size of the space means that natural reflections no longer help to improve the intelligibility of speech. Instead, the electroacoustic installation has to ensure high-quality acoustics for many kinds of concerts, plays and shows in addition to congresses and party conferences (Fig. 4).

Teleconferences
Rooms for teleconferences (video and/or audio) call for very special care to be exercised in the design to take account of the interaction between room, room acoustics and electroacoustics.

In such a conference, the sound is picked up by microphones in the room and transmitted to another place. While listening in the same room, the human ear is able to process directional information and cognitively "screen off" noises and disturbing reverberation to a certain extent. The expression "cocktail party effect" has been coined for this. However, microphones cannot accomplish this separation even if they have a distinct directional characteristic.

Background noise level
The background noise level in the room should be so low that communication is not impaired. It should therefore lie below 40 dB(A) whenever possible, better still below 35 dB(A).

Nowadays, the background noise level is very often raised due to the presence of media equipment, particularly video projectors, which can lead to a serious

reduction in intelligibility. Even if it is not always perceived consciously, such a situation requires extra concentration at least. For this reason, a maximum sound power level should be specified for media devices at the design stage by the person(s) responsible for planning the media installation.

Sound insulation
The sound insulation required between adjacent conference rooms or lecture theatres depends on their sizes and functions. A value of $R'_w \geq 45$ dB should be regarded as a minimum for small to medium-sized rooms without electroacoustic amplification. If the standards of DIN 4109 for educational establishments (and that

includes university lecture theatres) are taken as a guide, the figure for the minimum sound insulation is very similar, namely $R'_w \geq 47$ dB.

In order to achieve maximum freedom from disturbing effects when using sound reinforcement systems, however, a higher sound reduction index in the region of $R'_w \geq 57$–62 dB is necessary, which can sometimes be achieved with single-leaf reinforced concrete walls, but may well require a double-leaf construction. Where the rooms can be divided by means of movable room dividers to increase the flexibility of usage, then only with very careful design and construction is it possible to achieve sound reduction indexes

of about $R'_w \approx 45$ dB. If significantly better sound insulation is required, pairs of movable room dividers will be necessary, with an intervening cavity at least 1 m wide. Room-dividing screens that slide vertically can also achieve high sound insulation values.
To reduce impact sound, floating screeds are standard in university buildings, assembly halls and conference centres.

1 Typical distribution of reflective and absorbent surfaces in a lecture theatre
  a Section
  b Ceiling
  1 Reflective in the frequency range 250–2000 Hz
  2 Absorbent
2 Library and lecture theatre building, Weimar, 2005, meck architekten, Andreas Meck, Stephan Köppel
3 Plenary chamber of the German Bundestag in the former Reichstag building, Berlin, 1999, Sir Norman Foster
The volume of the plenary chamber is approx. 30 000 m³. With such an interior volume, it would be impossible to understand speech without electroacoustic assistance. The enormous volume per person of almost 30 m³ (MPs plus visitors) also required a considerable area of absorbent surfaces. The design of the stone and glass surfaces higher up the walls presented a particular challenge which was solved with a raised absorbent flooring system, absorbent stretch textile coverings behind the Presidium, absorbent ceilings and, last but not least, a customised electroacoustic system. In addition, the areas of glass below the dome can be acoustically deactivated by way of absorbent roller blinds.
4 National Convention Center, Hanoi, 2006, von Gerkan, Marg & Partner
In the large 3700-seat hall of the NCC Hanoi, large areas of the walls are covered with perforated wooden panels and absorbers behind stretch textile coverings. Part of the ceiling was also designed with perforated wooden panels. The space below the gallery can be shut off with a highly sound-insulating vertically sliding room divider.

Zollverein School of Management and Design in Essen

Architects: SANAA, Tokyo
Böll, Essen
Acoustics
consultants: Müller-BBM, Planegg
Completed: 2006

Following almost four years of intensive design and construction work, the new building for the Zollverein School of Management and Design was opened in July 2006 at the former Zollverein Colliery, a UNESCO World Heritage Site. The Zollverein School offers teaching and research courses to an international public which combine management and design skills in a new way.

The cube-shaped building measuring about 34 x 34 x 34 m houses an auditorium on the ground floor, a 9 m high design studio on the first floor and seminar rooms plus offices on the second and third floors. The unconventional architecture rendered special acoustic solutions necessary. For instance, sound-absorbent surfaces are absent from the ceiling of the 200-seat auditorium (see detail drawings) and practically absent from the walls. Instead, the floor was acoustically activated by laying a carpet (suitable for displacement ventilation and with an appropriate flow resistance) on perforated, raised floor panels. At the bottom of the walls, the carpet is affixed to a perforated plate concealing an attenuated void. This construction achieves broadband sound absorption encompassing all speech frequencies. The walls are of glass and the inner leaf is inclined inwards at an angle of 1° to prevent flutter echoes. The double-leaf glass facade not only houses the blackout blinds, but also insulates the room very effectively against the noise coming from the lively foyer with its cafeteria.

Glass and fair-face concrete surfaces dominate visually in this design, but the form of construction chosen nevertheless achieves the best acoustics for communication.

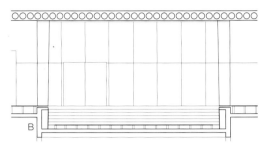

Sections through auditorium
scale 1:200

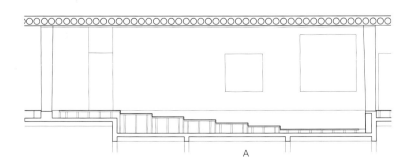

A Longitudinal section   scale 1:10
B Transverse section   scale 1:10

1 Flatweave carpet on 36 mm calcium silicate foam
  boards, proportion of holes 16 %
2 Wool felt, 400 g/m²
3 Longitudinal steel hollow section, 70 × 50 × 3 mm
4 Supporting framework of steel hollow sections,
  terraced, raised, 80 × 60 × 5 mm
5 Displacement ventilation outlet
6 Aluminium angle, 28 × 25 × 3 mm
7 Grille,
  anodised aluminium, concealed fixings
8 Double-leaf glass separating wall:
  laminated safety glass, 2 No. 8 mm toughened
  safety glass panes in aluminium channel section,
  40 × 35 × 3 mm
  565–610 mm cavity, with blackout blinds
  laminated safety glass, 2 No. 8 mm toughened
  safety glass panes in aluminium channel section,
  40 × 35 × 3 mm, inclined at 1°
9 Sill construction:
  3 mm sheet aluminium, perforated, proportion of
  holes 20 %, fleece backing
  steel hollow sections 40 × 40 × 3 mm,
  with 40 mm mineral wall insulation in between
  180 mm reinforced concrete
10 Reinforced concrete, trowelled smooth, flatweave
   carpet attached with adhesive
11 Acoustically attenuated void:
   flatweave carpet bonded to
   12.5 mm perforated plasterboard
   30 mm mineral wool acoustic insulation
12 Calcium silicate foam boards, 36 mm
13 Isolating tape
14 Cooling/ventilation

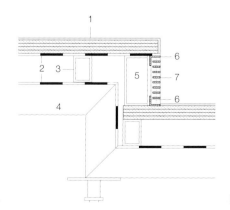

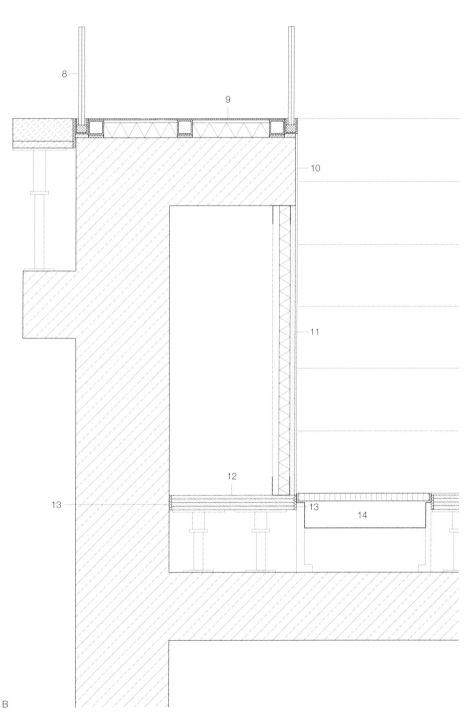

Small rooms for music

Compared to large halls for events, the acoustic design of smaller rooms for music is often neglected. The smaller rooms we are talking of here are rooms for teaching, practice and rehearsals in music schools and music colleges, but also the rehearsal, warm-up and conductor/soloist rooms in concert halls or opera houses.

In such rooms the acoustic feedback to the musicians is crucial. The size of such a (smaller) room alone results in totally different tonal relationships to those in large interior spaces intended for performances. Moreover, the sound insulation between neighbouring rooms plays an important role. As a rule, constructional measures are necessary in such cases – which far exceed the sound insulation standards typical in, for example, housing. This can represent a challenge to the architect, especially when converting existing buildings into music facilities. The acoustic planning tasks are very diverse and it makes sense to consider the acoustic aspects at the earliest possible stage, alongside the structural engineering, building services and lighting concepts. It is with this in mind that the important room and building acoustics requirements are summarised below. Two case studies are used to illustrate the broad spectrum of constructional acoustic measures.

Room acoustics

Many different factors affect the requirements placed on the room acoustics in rooms used for music tuition, practice and rehearsals. In rooms with too little damping, or in extreme cases no acoustic treatment at all, the loudness level is too high and, furthermore, such rooms tend to encourage imprecise articulation and tone formation. On the other hand, if the absorption surface area in the room is too large, the room is over-attenuated, which can sometimes curb the pleasure of playing and the motivation because correct intonation is made more difficult and the sonority of the instrument or voice is reduced.

Besides the instruments to be played and the level of competence of the musicians, the music to be practised and the long-term health effects for students and teachers must also be considered. A room that is ideal for professional practice may hamper the teaching of beginners, and a room suited to singing could be unbearable for a drummer.

1

1 Megaro Mousikis Concert Hall, Athens, 2007,
   A. N. Tombazis Associates
   In the ensemble rehearsal room (approx. 60 m²)
   the angled walls plus curving wall and ceiling elements contribute to a good sound mix. A basic
   attenuation is achieved via the ceiling, which is
   absorbent in some areas (perforated plasterboard
   for the flat sections). Sound-absorbent curtains
   allow the reflective character to be varied and the
   reverberation time to be adjusted between 0.6
   and 1.0 s.

Room volume

As the loudness level in a room essentially depends on the volume of the room, the size of the room should not be too small, particularly with louder musical instruments. The loudness level that ensues in a room that is too small is frequently impossible to keep within bearable limits, even with a high degree of attenuation. This represents a serious problem, specifically with respect to the long-term effects on the health of teachers and professional musicians. The recommended minimum size for individual lessons on loud instruments is approx. 40 m³ and at least 20–30 m³ per person for ensemble or orchestra rehearsal rooms; choir rehearsal rooms can be a little smaller (about 10–15 m³ per person). The larger the room, the more important it is to ensure an adequate ceiling height.

At the top end of the scale, a rehearsal room for a symphony orchestra with up to about 100 musicians will require a room with a floor area of at least 250 m² and a ceiling height of at least 8 m. With such rehearsal facilities, the mutual listening demands placed on the musicians also increase.

Room proportions and distribution of sound

Rooms with a square or rectangular plan form, or rather any proportions that promote the coincidence of natural frequencies, should be avoided (see "Room acoustics", p. 18). Such proportions promote disturbing irregularities in the bass range, which manifests itself in individual pitches that are either booming or do not resonate properly at all.

Placing walls at angles, possibly also the ceiling, – about 7° should be adequate – helps to avoid flutter echoes that distort the sound (Fig. 1). Furniture but also sound-scattering surfaces on the walls and ceiling make a decisive contribution to achieving a balanced sound mix.

Reverberation times

The recommended reverberation time for rooms for individual music tuition is about 0.6–0.8 s. Rooms for the teaching of children, e.g. singing, recorders, etc., should have a somewhat longer reverberation time, around 1 s, in order to increase the joy of playing. Reverberation times in this order of magnitude are preferred for aural training, too.

Contrasting with this, very high attenuation within the room and extensive absorbent surfaces on the ceiling and also the walls are indispensable in rooms used for band practices, drums and percussion. Adequate attenuation in the bass range is important for all these rooms. What this means is that a rise in the reverberation time for low frequencies should be avoided.

Acoustic interior design

Some basic acoustic attenuation should always be provided, which can be achieved, for instance, with an absorbent ceiling over part of the room at least. The effect of such attenuation should be as consistent as possible for frequencies, i.e. also absorb the lower frequencies. It is also beneficial when absorbent surfaces and reflective surfaces – or rather those that exhibit a diffuse scatter behaviour – alternate. Wooden linings with gaps between the individual elements fixed in front of an absorbent backing material ensure the necessary absorption of low frequencies.

The extent of absorbent linings depends on the desired reverberation time and the use of the room. Large areas will be needed in rooms for drums, percussion or big-band rehearsals.

If practice rooms are simply lined with carpets or a flatweave material, the result will be a dull, booming character.

It is always advisable to plan for variable absorption measures, which allows individuals to adjust the acoustic effect of the room as required. Sound-absorbent curtains are not expensive; they should be fitted to one wall, or two adjacent walls (Fig. 1).

More elaborate measures for influencing the room acoustics are rotating elements, e.g. built of wood, which are absorbent on one side and reflective on the other, or electrically operated roller blinds made from a sound-absorbent fabric or similar material.

Building acoustics

Adequate sound insulation between neighbouring music rooms is necessary to ensure essentially undisturbed simultaneous usage. Only very rarely does the specification stipulate that music played in one room should be inaudible in the next. It is more usual for the desired sound insulation to depend on the demands, the feasibility of the constructional measures and the budget. In existing buildings, the fabric of the building is very important from the acoustic viewpoint. The starting point for planning the building acoustics is the definition of a suitable sound insulation standard, i.e. the requirements placed on the weighted sound reduction index R'_w and normalised impact sound level $L'_{n,w}$. There are no standards or directives applicable here. Tab. T1 lists guidance values which are based on the experience of Müller-BBM gained in numerous projects. If we compare these figures with those typical in housing, we can see that the requirements are at least 10 dB higher, which is equivalent to halving the loudness level. Especially high demands (typically sound insulation category C) will almost certainly be required at music colleges or any other new music facilities designed to a high standard. With sound insulation category A, some interference with the function of the room is to be expected. Sound insulation category B is feasible in

1a, b   Bavarian State Opera, Munich, 2003,
Gewers, Kühn & Kühn, with Atelier Achatz
Architekten
In the conductor's room in the rehearsal building,
large-format decorative acoustic panels compensate for the generous expanse of glazing. The
highly sound-insulating coupled windows ensure
the quiet atmosphere which is vital.

T1: Guidance values for different sound insulation categories for music rooms (empirical values, Müller-BBM)

Component	Sound insulation category		
	A	**B**	**C**
Walls and suspended floors between-music rooms or between music rooms and other rooms in the same occupancy (e.g. offices)	$R'_w \geq 57$ dB $L'_{n,w} \leq 46$ dB	≥ 62 dB ≤ 38 dB	≥ 72 dB ≤ 28 dB
Walls and suspended floors between especially loud music rooms (e.g. drums, pop music, big-band, orchestra) and other music rooms or other rooms in the same occupancy (e.g. offices)	$R'_w \geq 62$ dB $L'_{n,w} \leq 38$ dB	≥ 72 dB ≤ 28 dB	≥ 82 dB ≤ 18 dB
Walls between music rooms and corridors	$R'_w \geq 47$ dB	≥ 52 dB	≥ 62 dB
Walls between especially loud music rooms and corridors	$R'_w \geq 57$ dB	≥ 62 dB	≥ 67 dB
Doors between music rooms and corridors	$R_w \geq 27$ dB	≥ 37 dB	≥ 45 dB
Doors especially loud music rooms and corridors	$R_w \geq 37$ dB	≥ 45 dB	≥ 52 dB

existing buildings with an appropriate underlying construction and careful planning.

Specifying requirements in terms of figures is easy, achieving these with construction is difficult. For this reason, the acoustic requirements must be discussed at an early stage of the planning and be taken into account in the loadbearing construction, the spaces (depth, thickness) allocated to floor and wall constructions, and, if applicable, the routing of ducts and services. Oversimplifying, we can say that for a sound insulation standard in the region of $R'_w \geq 62$ dB and $L'_{n,w} \leq 38$ dB, the space for floor and wall constructions will typically need to be about 250–350 mm. For $R'_w \geq 72$ dB and $L'_{n,w} \leq 28$ dB, about 350–500 mm will almost certainly be required. And that doesn't even include any lining (e.g. sound-absorbent materials) in the room itself!

Reinforced concrete, heavyweight masonry and/or non-rigid dry constructions using acoustic plasterboard or gypsum fibreboard are suitable materials for good acoustics. But separate rooms within the enclosing walls are frequently necessary,

i.e. acoustically decoupled walls, floor and ceiling are built inside the actual room itself (so-called room-in-room construction). Dry construction methods are suitable here, but also heavyweight materials, depending on the situation. Adequate sound insulation, especially for low frequencies, is important here, which means that the resonant frequency of such a double-leaf construction should never be higher than about 40 Hz. From this we can conclude that the depth of construction of such an inner shell must be approx. 150 mm at least. In very demanding situations, considerably lower resonant frequencies (< 20 Hz) may need to be achieved, which generally means building a heavyweight inner shell.

The potential for disturbance can be further reduced by ensuring that the layout of the rooms within the building aids the passive noise control. For example, especially loud drums or band practice rooms should be housed in a basement or semi-basement whenever possible.

Insulation against external noise

Rooms for music tuition in music schools and colleges should achieve the same level of sound insulation against external noise as that in classrooms in schools, i.e. a maximum level of 35 dB(A). But this maximum level should be reduced to about 25 dB(A) or less for high-quality rehearsal rooms and classrooms. Even higher demands apply when the rooms are to be used for high-quality recordings. In the case of a high level of external noise, or a demanding acoustics specification, mechanical ventilation will be necessary in addition to very good sound-insulating, coupled windows. Ideally, fresh air should be introduced into the room from the corridor via sound-insulated grilles (e.g. with baffles) because otherwise the sound insulation will be impaired.

One should not underestimate the problems that can occur due to sound emissions from music teaching facilities into the surrounding neighbourhood! The rule here is to apply the relevant provisions of the Technical Rules for Protection Against Noise (*TA Lärm*). Potential conflicts can be expected when the rooms are ventilated naturally.

Music school in Grünwald

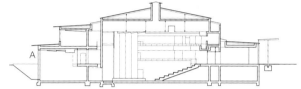

Architects: Peter Biedermann,
                   Rupprecht Biedermann, Munich
Acoustics
consultants: Müller-BBM, Planegg
Completed: 2002

The music school in Grünwald is a building with the highest acoustic demands.
A hall for the performance of chamber
music with seats for more than 300 – the
August Everding Hall – is located in the
centre of the building. Ranged around
this hall, and separated by corridors, are
the rooms of the music school – various
rooms for tuition, practice and rehearsals,
plus offices, spread over three floors.
Acoustic aspects were taken into account
right from the beginning of the planning
work for this building. For example, there
are no loud rooms directly adjacent to the
hall for chamber music performances. All
the elements – floors, walls, etc. – separating practice rooms are of such a construction that all rooms can be used
simultaneously without causing disturbances in adjacent rooms. Masonry and
independent wall linings on resilient bearing strips were required for some especially loud rooms, e.g. for drums, percussion and big-band rehearsals.
The circular plan shape of the building
automatically results in walls that fan out,
thus rendering unnecessary any wall linings to counteract the effects of parallel
walls.
All tuition, practice and rehearsal rooms
are equipped with variable acoustic
measures (curtains) so that the acoustic
conditions can be varied.

The rooms arranged in a ring around the concert hall
in conjunction with the heavyweight reinforced concrete
construction offer a very good initial situation right
from the loadbearing structure level. Depending on
usage, the high level of sound insulation required is
achieved with double-stud dry wall constructions plus
inner leaves of lightweight or heavyweight masonry.

Section and plans
scale 1:750

1 Music education
   for young children
2 Lobby
3 Big-band

4 Classroom
5 Green room
6 Artists' entrance
7 Office
8 Foyer
9 Concert hall
10 Control room

aa

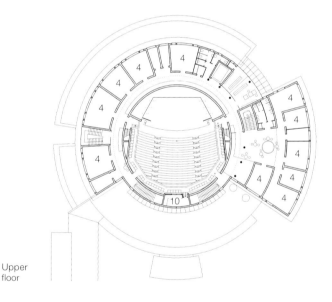

Upper
floor

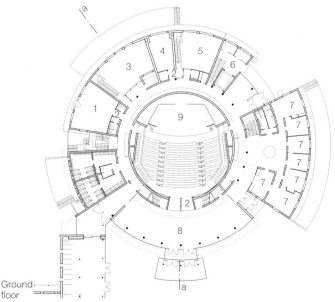

Ground
floor

Case study – small rooms for music
Music school in Grünwald

1 Large rehearsal room. To avoid flutter echoes between ceiling and wood-block flooring, the central part of the ceiling is divided into angled, overlapping sections. The perforations in the lower ceiling sections around the sides of the room contribute to the attenuation within the room.

2 Heavyweight independent inner leaf during construction. The rigorous decoupling of the leaf from the structure behind is critical for its acoustic effectiveness.
  a Steel framing
  b Infill panels of heavyweight clay bricks
  c Resilient fixing at the top of the leaf
  d Resilient bearing strip at the bottom of the leaf

Detail section   scale 1:10

Sketch showing the principle of the acoustics of the "heavyweight room-within-a-room" for the big-band practice room, with coupled window

1 Resilient fixing
  at top of inner leaf
  with rubber-metal elements
2 Permanently elastic seal
3 Suspended ceiling construction:
  80 mm flexible hanger
  60 mm mineral-fibre insulation
  2 No. 12.5 mm plasterboard, $m' \geq 20$ kg/m²

4 Coupled window:
  double glazing, 4 mm + 16 mm cavity + 6 mm
  sound-absorbent reveal ≥ 200 mm
  single glazing, 8 mm laminated glass
5 Mineral-fibre perimeter insulation
6 Resilient bearing strip,
  25 mm polyurethane elastomer
7 Floor construction:
  wood-block floor finish
  70 mm screed, separating layer
  30 mm mineral-fibre impact sound insulation,
  dynamic stiffness ≤ 20 MN/m³
  35 mm wood-wool lightweight building boards
  250 mm reinforced concrete

1

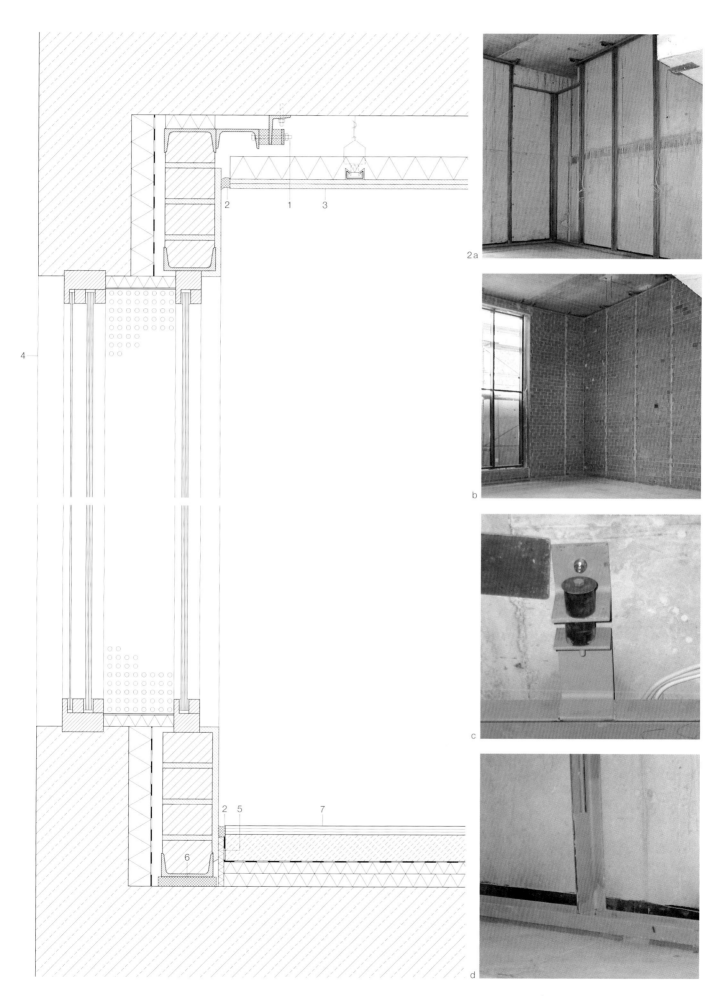

2a

b

c

d

Conversion of an officers' casino into a music school in Landshut

Architects: Nadler & Sperk, Landshut
Acoustics
consultants: Müller-BBM, Planegg
Completed: 2000

The intention behind modernising and converting the former officers' casino was to create a number of classrooms, a big-band room, a choir rehearsal room and rooms for the music education of young children, plus the associated offices. In addition, there are also rooms in the basement for a wind band, drums/percussion and folk music.

A beneficial aspect in the acoustic concept was the fact that the building had heavyweight external walls and some heavyweight internal walls that ensured good insulation against flanking transmissions. To improve the sound insulation between the practice rooms on the first floor and the rooms on the ground floor, the timber joist floor was upgraded structurally and acoustically. This was achieved by adding non-rigid ballast to the floor in the form of concrete flags with a minimal residual moisture content, which were bonded to the existing floor with bitumen,

and by laying a floating screed. The pugging was left in place because it is acoustically effective in this case. Additional, non-rigid soffit linings were provided in the rooms and classrooms below.

Some of the walls separating the practice rooms on the first floor were placed at an angle to avoid right-angles and parallel wall surfaces. Double-stud walls clad with plasterboard were used here. Non-rigid independent wall linings were built in front of the existing lightweight, solid internal walls.

Good sound insulation between the rooms and the corridors was achieved by using pairs of doors.

In order to avoid disturbing reverberation, a perforated plasterboard suspended ceiling, with an absorbent backing over 60% of the area, was erected in the rooms in addition to the direct measures to improve the sound insulation of the structural floors. One wall in each room is

fitted with absorbent curtains so that the room acoustics can be regulated.

A new reinforced concrete slab was built over the basement, where the facilities include practice rooms for big-band and drums. These rooms, too, were provided with extensive absorbent soffit linings plus sound-insulating independent wall linings and suspended ceilings. Owing to the existing situation, the resulting ceiling height in the basement is only just acceptable; lowering the basement floor would have led to excessive costs.

The greatest change to the existing building was the conversion of the former foyer into a 100-seat concert hall. The intermediate floor was removed in order to achieve the necessary room volume and hence acoustic resonance. A new gallery on three sides provides space for additional seating.

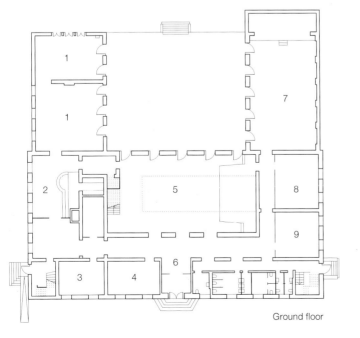

Ground floor

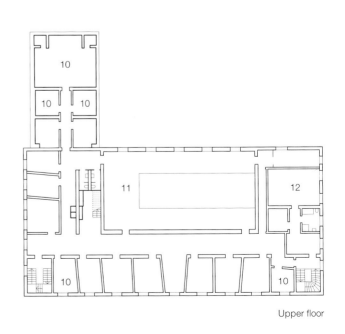

Upper floor

Plans   scale 1:500
Detail section   scale 1:10

Double-stud walls in dry construction plus an
acoustic upgrade for the timber joist floor ensure
the necessary sound insulation between the
practice rooms on the first floor and the ground
floor.

1   Music education for young children
2   Waiting zone for parents
3   Music school director
4   Secretariat
5   Concert hall
6   Foyer
7   Rehearsal room for choir and big-band
8   Media room
9   Staff room
10 Classroom
11 Gallery around concert hall
12 Apartment (existing)
13 Suspended ceiling construction:
    ≥ 60 mm resilient bars fixed to battens,
    with 40 mm mineral-fibre insulation in between
    2 No. 12.5 mm plasterboard, $m' ≥ 20$ kg/m²
14 Permanently elastic seal
15 Wall construction:
    3 No. 12.5 mm plasterboard, $m' ≥ 30$ kg/m²
    75 mm channel studs,
    with 60 mm mineral-fibre insulation in between
10 mm cavity
75 mm channel studs,
with 60 mm mineral-fibre insulation in between
3 No. 12.5 mm plasterboard, $m' ≥ 30$ kg/m²
16 Perimeter insulation
17 Floor construction:
    ≥ 25 mm asphalt on separating sheeting
    20 mm impact sound insulation
    concrete flags, bonded, $m' ≥ 120$ kg/m²
    particleboard, screwed down, joints filled
    timber joist floor (existing),
    with pugging in between
    ≥ 150 mm flexible hanger,
    with 40 mm mineral-fibre insulation in between
    2 No. 12.5 mm plasterboard, $m' ≥ 20$ kg/m²

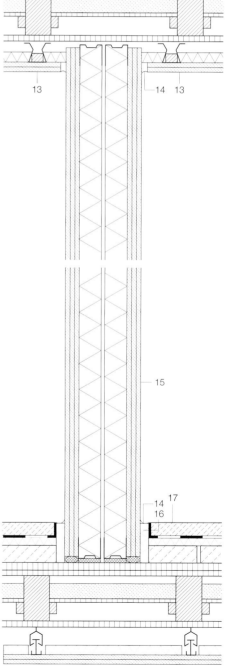

Rooms for sound

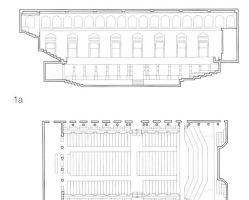

1a

b

c

When it comes to large venues for artistic and cultural events involving music, dance and/or drama, this is where acoustics skills become an art; at least that's the opinion often expressed. The discussions concerning the acoustics of such interior spaces are therefore correspondingly diverse and multifaceted.

The assessment as to what constitutes "good acoustics" depends on many factors. Besides scientifically measurable variables, influences such as the expectations of the listener or the atmosphere of the event also play a role.

Even if the boundaries between art, experience and science do tend to become blurred in the design of rooms for sound, certain relationships between acoustics and architectural design must be observed and implemented if the acoustics are going to be a success.

The following sections will describe important acoustic design aspects for rooms for sound and offer advice for their architectural design by way of examples. In doing so, a distinction will be made between room categories that require different acoustic treatment because of their function and size. These include dedicated premises for classical music, ballet, opera and drama. Buildings that frequently have to combine the aforementioned uses (speech and music), e.g. municipal theatres, civic centres, community centres and parish halls, are also discussed. It will also be shown how industrial buildings can be turned into acoustically successful performing arts centres and which acoustic attributes are required in venues for jazz and pop music.
The final section investigates variable room acoustics in performing arts venues and the topics of noise control and building acoustics.

Concert halls for classical music

The spectrum of concert halls for classical music ranges from relatively small halls specifically intended for soloists or chamber music right up to large halls for symphony orchestras with more than 100 musicians. Audience sizes vary accordingly – from less than 100 right up to 2000, in some cases even more.
In all these halls, it is musical performances without electroacoustic amplification that represent the primary goal, and "good acoustics" are essential for performances of classical music.

Room volume and reverberation time
In concert halls for classical music, the shape of the room, the volume of the room and the characteristics of the enclosing surfaces must guarantee a good sound mix and blending of the musical tones and hence a homogeneous orchestral sound. To achieve this, a much longer reverberation time is required than that normally found in a room for speech. For instance, in large concert halls for symphonic music, reverberation times of 1.8 – 2.0 s (with audience) are advantageous for the medium frequencies. The reverberation time should tend to increase noticeably as we approach the lower frequencies (about 1.3–1.5 times the reverberation time of the medium frequencies). Excessively long reverberation times should be avoided because they lead to a diffuse, undifferentiated overall sound.

Ideal room volumes for large halls for symphony orchestras lie in the range between 15 000 and 20 000 m³. The sounds of large orchestral forces, like those common among the early romantic composers (e.g. Dvořák, Tchaikovsky, Bruckner, Mahler, etc.), can then unfold adequately. This should not really surprise us; for such works were written for halls with such proportions. On the other

hand, a room with a volume exceeding about 25 000 m³ represents a difficult acoustic for many orchestral works, especially in terms of sonority and loudness.

In order to achieve the desired reverberation times, a volume of about 10–12 m³ per person is an important prerequisite.

Room forms
The rectangular room, acknowledged as acoustically good and often found in historic buildings, has proved to be a very good form for large symphony orchestras. Ideally, with a room about 19–22 m wide, or a little wider on occasions, this geometry leads to the audience being well supplied with intense early reflections from the side walls, which ensures a spacious sound impression. The better this 3D feeling, the more a listener feels he or she is "surrounded by the music", subjectively enlarging the source of the sound. In recent decades in particular, the three-dimensional effect has gained in importance for assessing the acoustic quality of concert halls.

The ceilings in such halls are generally horizontal, but other forms are also possible. Critical here is the transmission of the sound energy to the listener within max. 80 ms after hearing the direct sound in order to enhance the clarity of the sound. Reflections from the ceiling of the hall are particularly helpful here because the sound does not glance off the audience, where they would be absorbed and weakened. Classical coffered ceilings or ceilings with curved sections have proved worthwhile. Such ceilings direct part of the sound energy to the listeners while the remaining sound energy is reflected diffusely in other directions, and therefore remains available for generating reverberations.

1 Musikverein Hall, Vienna, 1870,
  Theophil Ritter von Hansen
  The Musikverein Hall in Vienna is indisputably one
  of the best concert halls in the world.
  a Longitudinal section
  b Plan on stalls
  c View of interior
2 Berlin Philharmonic, 1963, Hans Scharoun
  The Berlin Philharmonic combines outstanding
  architecture and successful acoustic design.
  a View of interior
  b Plan
  c Longitudinal section

2a

Examples of such rectangular halls are those of the Musikverein in Vienna (Fig. 1) and the large hall in Amsterdam's Concertgebouw, both of which were built in the late 19th century and are regarded as among the best concert halls in the world. A ceiling height of about 17 m and a distance from the front of the stage to the back rows of about 40 m ensure the necessary volume. But even many newly constructed halls stick to the tried-and-tested shoebox form, e.g. the concert hall in the Culture & Convention Centre in Lucerne or the Essen Philharmonic, which is described at the end of this chapter.

The Berlin Philharmonic, world renowned for both its architecture and its outstanding acoustics, represents a totally different interpretation of an interior space (Fig. 2). Here, the orchestra is placed more in the centre of the space. The individual audience blocks are divided up by sound-reflecting wall panels which are positioned in such a way that they guarantee a good supply of early side-wall reflections for all listeners.

Compared to a rectangular hall, an interior modelled on the Berlin Philharmonic concept has the advantage that the comparatively central position of the stage places the audience somewhat closer to the orchestra on the whole, and seats to the side of or behind the orchestra open up new aural and visual perspectives. For many concert-goers, the nearness to the musicians is a very special experience, many conductors and musicians also appreciate the nearness of the audience, regarding this "intimacy" as positive. In the Berlin Philharmonic no member of the audience is further than
30 m from the stage, despite the fact that this hall seats around 2200.

In halls with balconies set back at the sides and a not excessively wide stalls area, listeners can enjoy the very best

acoustic conditions. The role models here are the Neue Gewandhaus in Leipzig and the concert hall in the Megaro Mousikis Arts Centre in Athens (p. 92, Fig. 1). These examples show that various fundamentally different concepts can lead to successful acoustics in the building of concert halls.

In principle, the acoustics of the "more liberal" forms are less easy to master than those of the rectangular hall. But if the constructive cooperation between architect and acoustics consultant succeeds in implementing the acoustic requirements in an architecturally satisfying way, excellent, architecturally appealing concert halls are the result.

This should not be misunderstood as implying that a room of any shape can be turned into an excellent concert hall. For example, halls based on circular or elliptical plan forms, or those with spherical

forms or domed ceilings, can vary from acoustically critical to totally unsuitable.

Audience layout
Raking (i.e. terracing or sloping) the rows of seats for the audience is advantageous because it improves the sightlines and hence also the direct sound link. Whereas a relatively steep rake is normally chosen for theatres to achieve optimum sightlines, in a concert hall a somewhat shallower rake is preferred, partly in order to avoid reducing the volume of the room unnecessarily.

The audience in a concert hall is essential for the sound absorption and hence – for a given volume – is responsible for the reverberation time that can be achieved in the end. The rule here is that the sound absorption of persons decreases with the distance between the rows and the width of the seats. In new concert halls, the dis-

b

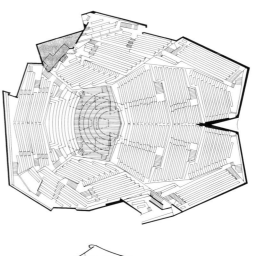

c

1a   b

tance between the rows of seats is generally 0.95 m and the seats are 0.55 m wide, but in historical buildings the rows are often placed much closer together. The sound absorption is particularly high in the case of small-format seating blocks because there are more people sitting adjacent to the aisles, thus presenting a larger total surface area for the absorption of the sound. This is an effect that should be avoided.

Materials

Apart from the seating, all the other surfaces in a concert hall for classical music are generally acoustically reflective. This means that solid surfaces with a high weight per unit area are required in order to achieve good reflection of the sound, also for low frequencies. The weights of wall linings and suspended ceilings are frequently in the region of 30–50 kg/m² (excluding the supporting construction). In principle, many different acoustically reflective materials can be used. It is frequently said that only wooden finishes produce good acoustics in concert halls, but this notion belongs to the realm of legend. In the Musikverein Hall in Vienna for example, just 15% of the wall and soffit linings are of wood – the rest is mainly plaster (p. 90, Fig. 1). Plasterboard and fairface concrete are good choices, in addition to wood and stone.

However, materials that exhibit a low internal damping, e.g. sheet steel or gypsum panels, should be treated with caution because they can lead to undesirable resonances in certain frequency bands. Furthermore, the subjective effects of materials and colours on the artists should not be ignored. If seemingly cold materials or surfaces mean that the performers do not feel relaxed, then that can certainly have an effect on the final sound.

Seating

Generally speaking, the acoustic properties of an empty hall or a poorly attended performance should not deviate too severely from those of the fully occupied condition. This means that the sound absorption of an unoccupied seat must be similar to the absorption of seat plus person when the seat is occupied. The design and materials of the seats must

therefore take acoustic aspects into account. Where the seats fold, the underside should be covered with an absorbent material. To avoid unnecessary absorption, seat surfaces not covered during use should be designed to reflect the sound; one suitable material for this is plywood. The rear of the back of the seat is one such surface. The upholstery to the back of the seat should not project beyond shoulder height.

Stage acoustics

The acoustic conditions on the stage, i.e. the area where the orchestra performs, are also given careful consideration these days. In the first place, the surrounding surfaces should provide a balanced distribution of reflections for the orchestra and therefore help the musicians to hear each other. Where the height of the ceiling above the stage exceeds about 15 m, reflective

1 Megaro Mousikis, Athens, 1992, Scroubelos et al. This concert hall in the heart of Athens has excellent acoustics, partly due to the reduced width at stalls level and the "sculpted" walls and ceiling. One special feature is that through the elaborate stage machinery, the hall can be altered to offer suitable settings for congresses, ballet, even opera, without having to make any compromises in the acoustics for concerts of classical music.
    a Hall changed into opera house, view from stage
    b Hall changed into opera house, view from auditorium
    c Hall with concert stage and organ   c

2

panels (individual elements or one large element) are often used these days. Such panels are typically suspended at a height of 8–12 m above the stage. Some can be adjusted to suit different orchestra sizes and seating arrangements.

The dimensions and layouts of stages for orchestras in different halls can vary greatly even though they have been designed for the same function! Besides the specific geometrical situation of the interior space, the traditions and experiences of the resident orchestra also play a role. If the stage is too large, each musician is allowed more "freedom of expression" and the subjectively perceived loudness is diminished, but the ensemble playing among the musicians is more difficult, and the balance of sound can suffer as a result.
As a rule, when planning for a large symphony orchestra these days, the width of the stage should be about 18–20 m, the depth about 12–13 m. An area at the back about 6–8 m deep should be added to this for choirs, but this area may also be used for additional audience seating when there is no choir on stage. It is normal for the stage (at least the front part thereof) to be 0.9–1 m above the level of the first row of seats.

The success of the stage layout at the Berlin Philharmonic has led to this design becoming very popular, i.e. the orchestra sitting in a semicircle around the conductor. In addition, the terracing on the stage can be varied, in the simplest case by way of demountable platform elements, or more elaborate and permanent mechanically operated stage lift systems. A stepped arrangement helps the sounds produced by different groups of instruments to be heard properly throughout the hall, which of course has a positive influence on the balance of sound.

Other uses and sound reinforcement systems
Concert halls, which are designed for non-amplified classical music, often have to be used for other functions purely for economic reasons. The range of uses ranges from pop music and shows to AGMs and conferences, and calls for electroacoustic amplification specially suited to each type of use and the necessary loudspeakers, which for appearances are preferably concealed.
However, the acoustic conditions in a concert hall primarily designed for classical music are hardly appropriate for these other uses because the enclosing surfaces optimised for reflecting the sound of an orchestra inevitably reflect the sound emitted by the loudspeakers as well, resulting in a diffuse, imprecise sound mix. With long reverberation times in the low-frequency range, the bass lines so important to pop music sound "uneasy". The only way to achieve a satisfactory solution is to use the very best sound reinforcement systems and loudspeakers with precise directional characteristics which direct the sound at the audience and therefore excite the hall as little as possible.
In addition, the loudspeakers should be positioned according to acoustic criteria so that correct acoustic localisation of the source is possible.

For these reasons it is advisable to consider the architectural effects of high-quality amplification and loudspeakers during the preliminary planning work and integrate these into the architectural concept.

Halls for chamber music
Large concert halls for symphony orchestras are often supplemented by smaller halls for chamber music, i.e. soloists and chamber music ensembles, but also smaller orchestras.

In halls for chamber music, clarity, the intimacy of the sound and the intelligibility of individual voices or instruments are very important factors, whereas the spaciousness of the acoustics is not as important as in large concert halls.

In order to achieve the desired clarity but at the same time a good sound mix, shorter reverberation times in the range 1.3–1.6 s are best. Such reverberation times can be accomplished with a volume of 7–10 m^3 per person.
Typical audience capacities lie in the region of 200 to 800 seats, which means a maximum room volume of about 8000 m^3. Other than this, the design criteria for large concert halls apply here as usual.

2   Franz Liszt Hall, Raiding, 2006,
    Atelier Kempe Thill
    A totally wooden concert hall for audiences of up to 600. An adequate room volume and the hardly noticeable convex curvature to the heavyweight panelled ceiling and walls lend the hall an excellent acoustic for chamber music, piano recitals and smaller orchestral works.

1a   b   2

Venues for ballet and opera houses

In terms of the room acoustics, the situation in an opera house is even more complex than that in a concert hall. After all, two types of performance are combined here. Besides the clarity of the singers and soloists, a good sound mix from the orchestra plus a good balance between orchestra and singers are all vital aspects.

The basic, historical form of the opera house is without doubt the "horseshoe". During the 18th century the horseshoe-shaped multi-balcony theatre with boxes for the nobles and upper classes started to spread across the whole of Europe, starting in Italy. The form is probably attributable to events held in palace courtyards or town squares, where performances were also watched from the windows of the surrounding buildings. Besides the desire to keep the classes separate, the balconied theatre also offered opportunities to watch not only the events unfolding on the stage, but also the other guests! Over the course of time, the number of boxes decreased in the newer opera houses, indeed, were even completely abolished in some, resulting in theatres with continuous balconies. But the horseshoe or a similar basic plan form was retained. Examples of this are the National Theatre in Munich and the Semperoper in Dresden.

A new development in the building of opera houses began with the Bayreuth Festspiel-haus, which Wagner had built for the performance of his works. In this theatre, the seating is laid out so that all members of the audience watch the events on the stage. The extremely deep orchestra pit, and the recesses in the side walls and the pillars are intended to accentuate the special mystical sound of Wagner's operas.

Apart from the Prince Regent Theatre in Munich, the design of the Bayreuth Festspielhaus was not copied elsewhere. However, it did inspire a new opera house design with a "more democratic" layout of the seats. Examples of such designs can be found in the Deutsche Oper in Berlin, the Opéra de la Bastille in Paris or the Small Festspielhaus in Salzburg, the "Kleines Haus für Mozart", which was completed in 2006, the 250th anniversary of Mozart's birth (Fig. 2).

Room volume and reverberation time
The reverberation times of famous opera houses – for the medium frequencies – lie between 1.1 s (La Scala, Milan) and 1.8 s (National Theatre, Munich). These days, the trend is towards even longer reverberation times (1.5 – 1.8 s).

This shows that *zeitgeist* and historical developments in music and drama plus the creation of new forms of performance can also have an influence on how the acoustics of a venue are gauged. For instance, in past centuries more emphasis was placed on the intelligibility of speech because a great number of new and unknown works were being performed. Nowadays, a greater fusion between the sound of the voices and the instruments is preferred in opera, which entails a slight decrease in the intelligibility of the words.

A typical room volume from the acoustics viewpoint would be less than 15 000 m^3 (excluding the stage) in a building designed for an audience of no more than 2000. Of course, there are far larger opera houses in use, e.g. the Opéra de la Bastille in Paris or the Metropolitan Opera in New York, which owing to their size, and despite best possible acoustic design, are really only suitable for the most powerful voices in the opera world.

Room form and ceiling design
With the aforementioned room forms, it would be desirable from the acoustics viewpoint for early reflections to assist the transmission of sound from the stage and the orchestra pit to the audience. To achieve that, a plan shape is required that does not fan out too severely in the front one-third of the space (approx. 15°).

In rooms with a concave plan shape (horseshoe, arc, etc.), the sound is sometimes focused to certain areas of the audience, even in historical buildings renowned for their acoustics.
Dividing up the wall surfaces and the edges of the balconies is helpful in such cases. The advantage of dividing the balconies into boxes for this room form is that the sound cannot travel along the curving wall by way of multiple reflections (whispering gallery effect).

The design of the ceiling is especially significant in an opera house. The reflections from the ceiling contribute to ensuring good intelligibility, definition and clarity of the sounds produced by the singers and the orchestra. To achieve this, a good supply of ceiling reflections should be

1 The new Opera La Fenice, Venice, 2003, Aldo Rossi
    Following a devastating fire in 1996, eight years later the rebuilt Opera La Fenice has regained its original splendour and once again can offer superb acoustics for audiences of up to 1000.
    a View of auditorium
    b Stage with large orchestra shell for performances of classical music
2 "Kleines Haus für Mozart", Salzburg, 2006, Holzbauer & Irresberger and Hermann & Valentiny
    A 1600-seat hall has been built on the site of the former Small Festspielhaus. The new auditorium could only be widened very marginally so it was shortened and made higher. But this has given Mozart a late gift in Salzburg: a compact building which thanks to its ideal internal dimensions assists the music and results in a transparent sound.

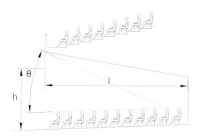

3   The ratio of balcony height h to balcony depth l
       should be max. 1:2 in an opera house; the angle θ
       should be min. 25°.
4   National Opera House, Helsinki, 2007,
       HKP Architects
       Ray geometry study for the conversion work. The
       proscenium ceiling was designed in such a way
       that much of the sound energy is reflected back
       into the orchestra pit.
       a Auditorium
       b Proscenium/orchestra pit
       c Stage

3

guaranteed for every seat, arriving within 80 ms of the direct sound. In the central balconies of many opera houses these tense early ceiling reflections are responsible for making the acoustic events appear closer than the visual impression would lead us to expect. Such seats are often the best from the acoustics viewpoint.

The design of the ceiling always involves conflicts between the requirements of natural acoustics and stage lighting. Access to the lights is achieved by way of high-level suspended gangways positioned at the points that are especially important for acoustic reflections. But a suitable, coordinated ceiling design usually results in a good solution.

Alternatively, in modern opera houses with less ornate ceilings it is possible to suspend the lighting bridges below the ceiling and to make them acoustically transparent or provide them with reflective panels. The latter is an excellent answer for good acoustics, as the Festspielhaus in Baden-Baden demonstrates.

Sightlines

In order that the movements and gestures of the performers are discernible, the distance from the last row of seats to the stage should not be more than about 30–35 m. Good visual conditions and good acoustic conditions are directly related, especially with respect to the supply of direct sound from the vocal soloists.

The sound of the orchestra emitted from the orchestra pit can, however, be very impressive in the gallery seats high up in the theatre. In the classical horseshoe-shaped balconied theatres such seats seldom have a direct view of the whole stage, but benefit from intensive ceiling reflections from the orchestra pit.

Balconies and boxes

Different acoustic conditions to those in the stalls prevail in boxes and below balconies. The shape and depth of the boxes plus the materials used are the main influences here.

To ensure that seats below cantilevering balconies enjoy a more or less consistent acoustic quality, the cantilever of the balcony in relation to the opening at the front should not be too large (Fig. 3).

Materials

As a rule, large sound-absorbent surfaces are avoided in opera houses in order to promote a longer reverberation time and ensure an exciting sound. In most cases the attenuation necessary is achieved by the audience itself and/or the seat upholstery. The same acoustic design principles apply here as for the seats in a concert hall.

Proscenium

The proscenium, i.e. the apron in front of the curtain separating the auditorium from the stage, is particularly sensitive from the acoustics viewpoint because it places reflective surfaces in the vicinity of the sound sources.

The proscenium ceiling and walls must be designed in such a way that the sound of the orchestra is also reflected back into the orchestra pit in order to improve the aural contact between the musicians. At the same time, the proscenium should also assist in transmitting the sound from the stage to the auditorium, which helps to improve the balance between orchestra and singers. Careful coordination between stage lighting, acoustics and architecture is necessary for the proscenium as well if good results are to be achieved (Fig. 4).

Orchestra pit

In an opera house the orchestra plays in the so-called orchestra pit so that the audience's view of the stage is not obstructed. This screening also helps to create a balanced sound.

The size of the orchestra pit depends on the maximum number of musicians to be accommodated. On average, each musician requires about 1.5 m². Typical sizes lie between 90 and 130 m²; the aspect ratio, i.e. ratio of length to width, should be in the order of 1:2–1:2.5, e.g. 7 x 16 m.

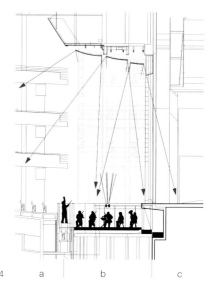

4   a   b   c

Normally, the rear part of the orchestra pit is underneath the stage itself in order to reduce the distance between stage and audience. This also helps to counteract the dominance of louder groups of instruments, which are then placed beneath the overhanging stage. However, as the musicians below the stage are exposed to an even higher sound level than their colleagues elsewhere in the orchestra pit and the attenuation of individual groups of instruments may be unhelpful, depending on the piece being performed, the overhanging stage should not cover more than about one-quarter to one-third of the orchestra pit. In addition, the surfaces beneath the stage and the rear wall should be given some form of acoustic treatment.

In newer opera houses, the depth of the orchestra pit can usually be varied in order to suit the sound requirements of works from different music periods. For example, the clarity of a Baroque opera requires a smaller depth, but for late-Romantic operas, in which the blending of the instrumental sounds and a dynamic preference for the singers is important, a deeper orchestra pit is used.

Stage
The design of the scenery is not part of the architect's remit. But if the set designer ensures that the scenery is as reflective as possible (plywood instead of fabrics), the acoustic transmission quality from the stage to the auditorium can be improved.

Use for concerts
Opera houses are also frequently used for performing classical music. A so-called orchestra shell is erected for this, which improves both the aural contact between the musicians themselves and also the transmission of the orchestral sound into the auditorium. Besides the acoustic aspects of such orchestra enclo-

sures, the handling of the segments and their storage (space requirements) when not in use are important aspects to be considered.

Sound reinforcement and sound effects systems
Electroacoustics have, in the meantime, gained more and more of a foothold in opera houses, especially in the form of surround-sound systems which generate directional or moving sound effects in the auditorium. Such systems allow new artistic freedoms in the composition and production of new works. The numerous loudspeakers required can generally be integrated into the wall linings so that they do not have any detrimental effects on the interior design.

Quite often, the acoustic contact between stage and orchestra pit is assisted by a customised loudspeaker system, which is installed in the region of the apron or stage. This is in no way intended to be amplification for the audience, but instead serves purely to improve the contact between the performers on the stage and the musicians in the orchestra pit, to assist intonation and ensemble playing.

Musicals and shows
For musicals and shows, the singers and the orchestra are always assisted by some degree of electroacoustic amplification, which means that the natural acoustic quality of the building is less important than would otherwise be the case. Notwithstanding, it is still sensible to follow the basic acoustic design principles of an opera house (proscenium, reflective ceiling) in a theatre for musicals and shows as well. Shorter reverberation times, around 1.0–1.3 s, are, however, desirable, and that means the rear wall surfaces, possibly also parts of the side walls, should be designed as absorbent.

Care should be taken to avoid individual reflections, especially in large theatres where the desired, short reverberation time has been achieved. Owing to the long transmission paths and the lack of masking by reverberation, disturbing echoes cannot be ruled out.

Theatres for drama
In theatres for drama without musical accompaniments, the most important acoustic requirement is good speech intelligibility. This calls for, on the one hand, appropriate damping, or rather an appropriate reverberation time, and on the other, reflective surfaces to boost the loudness level of the actors' lines.

Reverberation times
Depending on the volume of the interior space, a reverberation time of 0.8–1.2 s is favourable for medium frequencies. A rise in the reverberation time for the low frequencies, which is very desirable for musical performances, should be avoided in venues for speech and drama, or at least reduced to a minimum.

Room volume
The volume should be about 4–6 m³ per person. The volume of the auditorium should not be greater than 5000 m³ in buildings used exclusively for drama where there is no electroacoustic amplification. In larger theatres it is almost impossible for the actors – without amplification – to speak loudly enough to fill the entire interior with their voices.

1 Refurbishment of the Kammerspiele, Munich, 2003, Gustav Peichl, Walter Achatz
The art nouveau building, protected by a preservation order, was extensively refurbished and restored between 2000 and 2003.

Room form and audience layout

Many different room forms satisfy the acoustics requirements for drama performances – provided the ceiling form helps to provide reflections and prevents any focusing of the sound, as can happen, for example, with concave surfaces. Reflections from the side walls, which are extremely important in a concert hall and not unimportant in an opera house, play only a minor role in the intelligibility of speech. Consequently, in theatres for drama even spaces that fan out severely are possible, provided good sightlines can be maintained. The distance from the front edge of the stage to the last row of seats should not exceed 25 m if the facial gestures of the actors are to remain discernible.

Distribution of absorbent and reflective surfaces

Early sound reflections, especially from the ceiling, should be encouraged, but intensive reflections that arrive more than 50 ms after the direct sound must be avoided. In the light of this, the ceiling should be designed to reflect and redirect useful reflections. Parts of the wall surfaces, and especially the rear wall, can generally be lined with sound-absorbent materials in order to create the necessary attenuation and prevent late sound reflections.

Upholstered seats help to reduce the reverberation during rehearsals of stage works as well.

Fly tower

As in an opera house, the acoustic situation in the fly tower, or rather on the stage itself, is determined by the (usually) very great volume of this part of the building. The volume of the fly tower can be similar to the volume of the auditorium, indeed, can even far exceed this. If absorbent backdrops and borders (fabrics masking the top of the stage as seen from the auditorium) are not stored (i.e. hung) here permanently, the fly tower will exhibit considerable reverberation, which will have an effect on the rest of the acoustics and will not help the speech intelligibility. For this reason, it is advantageous to provide permanent absorptive surfaces in the fly tower which guarantee a basic level of attenuation regardless of any scenery and props. The underside of the roof to the fly tower and the walkways can be designed as absorptive. Black-painted perforated sheet metal or dark-coloured wood-wool acoustic boards are frequently used.

Experimental theatre

Besides the traditional theatre with its proscenium stage, there are modern theatre forms that try to bring the action on the stage closer to the audience. A concept that has proved worthwhile is the so-called black box, in which the stage and the audience seating can be rearranged as required to suit each production, e.g. a catwalk or an arena. The acoustics of such interiors also benefit from the comparatively short distance between actors and audience. The direct sound alone is often good enough to ensure good intelligibility of speech. However, as an actor cannot face in all directions at once, additional ceiling reflections are important for ensuring that all members of the audience are supplied with sufficient reflected sounds at all times, even when an actor is facing the other way. Suspended reflectors or a reflective ceiling can be used to provide the necessary reflections. The other wall and ceiling areas can be lined with sound-absorbent materials in order to prevent disturbing reverberation.

Electroacoustics

Electroacoustic systems are often used in theatres for amplifying the actors' voices but also for generating sound effects (e.g. surround sound). The relatively high internal damping required for good speech intelligibility in a venue for drama basically also favours the use of electroacoustic amplification systems. When choosing loudspeakers and deciding on their positions, care should be taken to ensure that in addition to providing all areas of the auditorium with adequate sound, the necessary directional quality is achieved so that listeners can localise the source (generally one or more actors on the stage).

Municipal theatres, civic centres, community centres, parish halls

Whereas the types of hall described up to now in this chapter are essentially intended for specific functions and therefore their acoustic design can be optimised accordingly, municipal theatres, civic centres, community centres and parish halls, and school assembly halls, are mostly used for a wide range of different events. Besides various drama and music performances, such premises are just as likely to be used for conferences, exhibitions or festivals.

Consequently, these multi-purpose interiors frequently include acoustic design elements from the interiors described above, depending on the main types of usage. But owing to the varying requirements, it is difficult to formulate the room acoustics conditions that will lead to a more or less satisfactory acoustic result for every type of usage. As is shown below, this can be achieved in different ways.

High interior attenuation
One possible room acoustics concept consists of employing a relatively high degree of acoustic attenuation within the interior itself. This is essentially achieved by designing according to the criteria for a room for speech. The high interior attenuation also offers good acoustic conditions for festivals and amplified music. However, with such a hall design, the reduction in the acoustic quality for classical music, which requires a longer reverberation time, has to be accepted.

Compromise acoustic solution
Another solution is an acoustic design that represents a compromise between speech and music. The attenuation in the room in this case is selected so that the reverberation time lies in the middle between the comparatively short reverberation time required for speech and the long reverberation time needed for classical music. This compromise solution is often employed, especially in parish halls, community centres and civic centres. If good acoustics for classical music are important, it is essential to provide an adequate room volume, ideally at least approx. 7 m³ per person. It should also be remembered that the seating in such halls is variable. Lightly upholstered chairs help the acoustics, but the room should still not be too reverberant without any seating at all.

Acoustic variability
Where high acoustic demands are placed on all functions for which the building is used, then it is possible to alter the acoustics to suit each type of event. If this is achieved by way of acoustically variable surfaces, here again, an adequate room volume will be necessary. The possibilities of variable room acoustics are discussed in detail later in this chapter.

Stage, orchestra enclosure, orchestra pit
From municipal theatres and civic centres right down to smaller community centres, all usually have a proscenium stage with stage equipment to match the likely uses. The stage acoustics should also be adapted to suit the range of events. That could mean reflective surfaces around the stage for classical music concerts; ideally, reflective ceiling and wall panels should form an orchestra shell.

On the other hand, for presentations, but also electroacoustically amplified concerts, higher attenuation is desirable, which can be achieved by way of suspended textile elements, for instance. If, in addition, an orchestra pit is required, then the same design criteria apply as for the pit and proscenium in an opera house. The floor of the orchestra pit can usually be raised and lowered between stalls and stage level.
Customised solutions to suit the size of the project and the budget of the client can be designed to suit the range of uses envisaged.

Industrial arts venues
In recent years many former industrial or trade fair buildings have been converted into permanent or temporary venues for cultural events. Such structures offer an especially creative environment and performance options that are often not possible in standard concert halls or theatres.

Such buildings were never intended to be used as venues for the performing arts; the sizes of such spaces alone often throw up completely different acoustics issues to those experienced when planning conventional halls and theatres. In light of this, it must be established early on in such projects as to whether the existing conditions can be adjusted to create an interior space of sufficient size

and favourable proportions. Further aspects to be considered are the possibility of disturbing noise caused by rainfall or aircraft flying overhead, and how such noise can be kept within acceptable limits.

Given the right conditions, first-class performing arts venues can be realised with a comparatively modest budget. One good example is the temporary venue built for the 1997 International Lucerne Festival (Fig. 4). At the time, the new Culture & Convention Centre was still under construction and so a 1750-seat hall was built in a large warehouse. The temporary hall had roughly the proportions of a classical rectangular concert hall and provided excellent acoustics for the world-class orchestras and soloists.

1   Civic Centre, Tuttlingen, 2002,
    Heckmann Kristel and Jung Architekten
    Optimised ceiling form and wall reflectors ensure a natural acoustic. In addition, an electronic room acoustics system enables the acoustic quality of the interior to be adjusted "at the touch of a button". The hall normally seats 850, but this can be increased to 1100. A small hall (250 seats), a conference room, four seminar rooms and a small open-air stage round off this complex.
2   Civic Centre, Hausach, 2006, Lehman Architekten
    A former gymnasium has been converted into a civic centre with 600 seats. The overlapping ceiling sections ensure early reflections and the sculpted structure of the walls counteracts flutter echoes. Narrow flat panels around the edges of the ceiling prevent excessive reverberation and the front of the balcony is absorbent.
3   Parish Hall, Forstenried, 2001, Cornelius Tafel
    The ceiling height in this parish hall (floor area about 160 m²) exceeds 6 m, which ensures a good room volume for the acoustics. The lining to the ceiling is mainly reflective, but there is an absorbent backing along the sides and at the rear. This method prevents an excessively live acoustic without over-attenuating the room. The small stage is stored in a room (which can be used separately) behind the sliding red partition.

4 a   b

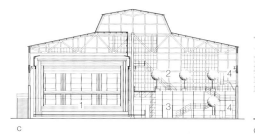

c

d

1	Concert hall	5	Cloakroom
2	Gallery	6	Box office
3	Foyer	7	Orchestra
4	Bar	8	Choir

In addition to the special atmosphere of such events, former industrial buildings provide chances for new, previously untried staging options, as Bochum's Centennial Hall demonstrates. This venue, known as the "assembly hall for art", stages many spectacular and first-class concerts and theatrical events. The huge volume has been tamed by the use of variable sound reflectors and an electronic room acoustics system.

4   Temporary concert hall, Von Moos Civic Centre,
    Emmenbrücke near Lucerne, Max Schmid
    The walls to the temporary hall were built from a
    concrete formwork system.
    a View from foyer
    b View of interior
    c Section through building showing concert hall
       enclosure
    d Longitudinal section through concert hall
5   Centennial Hall, Bochum, 2002,
    Petzinka Pink Architekten

5

Halls for jazz and pop music

Although jazz and pop music play important roles in our modern cultural lives, concerts of such music frequently take place in less than suitable acoustic surroundings. Classical music venues, sports centres, former industrial buildings and marquees are used, without any thought as to their acoustics. The result is that the quality of the sound suffers enormously. More recently, however, more attention has been given to providing good room acoustics, including by the operators of such venues.

The most important criterion is to ensure adequate attenuation of the sound within the room. To do this, the reverberation time should not exceed 1.0 s, indeed, shorter reverberation times are even better. However, in large arenas, reverberation times of up to 1.8 s are still acceptable. The reverberation time should not rise for the low frequencies, but instead should, if possible, drop slightly. Generally, large areas of sound-absorbent materials – effective over the widest possible range of frequencies – are required on the ceiling and walls. Perforated panels made of metal, wood or plasterboard, with a sound-absorbent backing, are suitable. Additional, special absorbers are required for absorbing low frequencies and these are often included in the form of plate absorbers.
Seating, at least permanent seating, is frequently absent in such venues, and cannot be included in the sound absorption calculation. The possible problems of individual reflective surfaces in heavily attenuated large interior spaces has already been mentioned in the section "Musicals and shows".
The electroacoustic amplification system represents an essential element at such a venue and must be perfectly matched to the room acoustics.

Absorbent walls around the stage are advantageous because they reduce the undesirable acoustic feedback from the monitor speakers into the auditorium. Such loudspeakers on the stage are necessary so that the musicians and vocalists can hear themselves and each other directly – essential for a good performance.

Variable and virtual room acoustics

In venues offering music, theatre and/or dance productions, multi-functional halls and concert halls used for other events as well, it is advisable to consider the possibilities of variable acoustics at the design stage.

Variable surfaces

The room acoustics conditions can be adjusted, for example, by providing acoustically variable surfaces, preferably on the walls. In the simplest case hanging curtains in front of an acoustically reflective wall or electrically operated roller blinds may be sufficient. Wall elements that pivot or fold to reveal absorptive surfaces or absorbers that can be stored in a ceiling void are other alternatives.
In order to achieve the desired effects, however, a large proportion of the total surface area must be acoustically variable, amounting to perhaps 40–80 m² per 1000 m³ of room volume, depending on the absorption required.
Crucial to the success of variable acoustic measures is their ease of use by the technical staff of subsequent users.

Reverberation chambers

In the past, concert halls were sometimes designed with reverberation chambers to achieve acoustic flexibility. These are rooms with hard walls which are placed around the hall and can be joined to or separated from it by doors. This solution enables the acoustically effective volume to be varied, e.g. increasing the reverber-

ation time for an organ concert. However, the subjective perception of this reverberation can be limited by the make-up (in terms of time and space) of the resulting sound field.
Absorbent curtains can be included in the reverberation chambers to reduce the resonance.
The Lucerne Culture & Convention Centre has reverberation chambers around its concert hall (Fig. 1). The total volume of the reverberation chambers here is a remarkable 6000 m³, i.e. equivalent to the volume of a full-size chamber music hall.

Electronic room acoustics

Another way of achieving variability is to use electroacoustics to influence the room acoustics. Such an approach can be appropriate for venues offering music, theatre and/or dance productions and multi-functional halls. Existing rooms with room acoustics shortcomings can also be upgraded with such a method.
Electronic room acoustics systems take the natural sound in the room, allow this to be reflected to a certain extent by virtually generated walls and thus adapt the reverberation and the reflection characteristics depending on the type of use. Expressed in more technical terms, the sound signals are picked up by microphones, changed with the help of signal processors and elaborate software and played back into the room via loudspeakers distributed evenly throughout the area. Diverse presets make it possible to call up different acoustic situations "at the touch of a button".

The latest systems can achieve a very natural aural impression which satisfies even demanding listeners and musicians. The resulting sound must, however, "harmonise" with the room. Excessive acoustic changes, e.g. turning a small community centre into a cathedral (acoustically,

1   Concert hall, Culture & Convention Centre, Lucerne, 1998, Jean Nouvel

that is!), are inevitably experienced as unnatural and can therefore only be regarded as artificial effects. Electronic room acoustics systems have little in common with conventional sound reinforcement systems and their design must take into account a room's natural acoustics.

Protecting against noise and vibrations

One of the most important quality attributes in rooms for sound is their ability to facilitate silence. This is the only way to ensure that the quietest passages of music or text are not lost in the background noise. In acoustically demanding venues, it is normal to specify a maximum of 25 dB(A) for disturbing background noise; requirements are also placed on the spectral composition of the disturbing noise level. Achieving a low level for any disturbing background noise calls for careful planning, which begins with the choice of location for the building and the design of the internal layout and continues right up to the specifications for passive noise control measures and building services.

Exterior noise situation and choice of location
When choosing a suitable location for a building in which artistic performances will take place, the exposure to exterior noise must be investigated so that the necessary noise control and sound insulation measures can be estimated and later designed. Besides the noise of road or rail traffic, the effects of other noise sources should not be ignored, e.g. church bells, the sirens of emergency vehicles, low-flying helicopters transporting patients to nearby hospitals, etc.

Vibrations and secondary airborne sound
One special subject is the vibrations caused by rail traffic, trams and underground trains where these are in the vicin-

ity of the site. The usually low-frequency disturbing noise is transmitted as structure-borne sound via the foundations into the building and re-radiated there in the form of so-called secondary airborne sound. Such sound propagation is often very difficult to deal with. Possible remedies are separate enclosures within the walls of the main building, elastic bearings for whole sections of the building or, alternatively, the track-bed, but all must be individually designed to suit the particular situation.

Planning the interior layout
Loud rooms, e.g. plant rooms, and other rooms for events or rehearsals, should not be positioned directly adjacent to a hall for performances. The same is true for lift shafts and also sanitary facilities, which should all be kept well away from the walls to any hall in which artistic events take place.
Acoustically beneficial is the placing of corridors, circulation zones and foyer areas around the hall to create an effective buffer. Furthermore, rooms grouped around the hall offer good protection against exterior noise and sometimes save the cost of elaborate sound-insulating wall constructions.

Construction of the building
A favourable interior layout is not normally sufficient on its own in order to achieve the necessary level of sound insulation for a hall for music or drama.
Generally, apart from plant rooms, floating screeds or, alternatively, resilient textile floor coverings will be required throughout the entire building in order to prevent the propagation of disturbing impact sound.

The walls to the hall itself must achieve a sufficient level of airborne sound insulation, which must be defined in each case.

A heavyweight form of construction using reinforced concrete is frequently advantageous for the acoustics. In new buildings, acoustically effective isolating joints can sometimes be a very effective way of separating loud parts of the building from other parts requiring protection. In some cases it may be necessary to build a completely separate room within the main room, with the inner room fully acoustically decoupled from the rest of the building. Access doors leading from loud foyers frequently make use of acoustic lobbies with pairs of doors. An effective separating joint in the screed is absolutely essential beneath such doors.

Noise of rainfall
One special consideration is the potentially disturbing noise of rainfall on roof constructions to halls or fly towers. If the roof is of a lightweight construction, measures must be taken to avoid the build-up and transmission of the noise of rainfall on the roof. A double-skin construction is one way of achieving this, with the two roof levels separated by soft thermal insulation to prevent the transmission of structure-borne sound. Stretching fine-gauge nets across the roof can also reduce the noise of rainfall. However, leaves and other debris must be removed regularly and the cost of this should not be underestimated.

Building services

In top-quality establishments (especially venues for classical music, opera houses and theatres offering high-class productions), the noise level due to building services should generally not exceed 25 dB(A). If the demands are not quite so high, e.g. venues for musicals, cabaret, etc., a value up to about 5 dB higher may be acceptable. It is usually advisable to specify frequency-related figures for the individual building trades (ventilation, lighting, etc.).

Especially low noise levels can be achieved with displacement ventilation systems in which the fresh air enters the room at a low velocity via floor outlets and the stale air is extracted via outlets in the ceiling. In addition, careful acoustic planning of the ventilation system and constructional measures are necessary, e.g. isolating bearings for the plant to prevent the transmission of structure-borne sound, sufficiently large duct cross-sections to guarantee low air velocities and hence minimise disturbing noises, and silencers at the right locations. The measures should be integrated into the planning work at an early stage because silencers, for example, require more space.

Stage machinery

Stage machinery and lighting used for and during performances cannot keep below the aforementioned maximum noise level of 25 dB(A), even when using the very latest technology. However, such machinery is usually used only briefly during performances and can then be incorporated into the production in such a way that, for example, a somewhat higher noise level occurs simultaneously with a louder passage of music and therefore is not heard by the audience.

A noise level of about 35 dB(A) is realistic for stage loft machinery. But for the stage floor machinery, e.g. stage lifts, revolving stages, etc., a noise level 5–10 dB higher than this must frequently be accepted. The maximum permissible noise level for stage machinery should be defined by the acoustics consultant in consultation with the stage designer and in the end should be specified in the tender for the stage equipment.

Any noise from stage lighting or projection systems can be particularly irritating because such equipment is often used for longer periods. In the meantime, manufacturers have responded and now supply sound level figures for their products, which can be taken into account when selecting equipment.

Even the normal house lights can cause acoustic problems if, for example, dimmer systems lead to buzzing noises, or sudden loud noises are emitted from lamps as they cool down, or music causes diffuser grilles to vibrate audibly.

Acoustic compatibility

When planning buildings for cultural events, it is generally necessary to investigate the effects on the local neighbourhood (see "Noise control in urban planning", p. 39). Conflicts can occur when the sound insulation to the building envelope, or parts thereof (e.g. glass facade), is only low and at the same time there are buildings nearby that require protection against excessive noise.

The sound immissions can be ascertained and evaluated by way of an acoustic compatibility study, which examines all noise emissions caused by events, deliveries, building services and car parking. If the permissible noise levels are exceeded, measures in, on and around the building will be necessary to compensate for this.

Opera house in Hangzhou

Architects: Carlos Ott Architects, Montevideo
Acoustics
consultants: Müller-BBM, Planegg
Completed: 2005

In 2005, a new arts centre was opened in Hangzhou, a city with 6 million inhabitants about 200 km from Shanghai. The centre has a large opera house (1600 seats), a concert hall (600 seats) and a multi-purpose hall (black box with 400 seats).

Great importance was attached to achieving perfect acoustics in the opera house, which had a direct effect on the design of the interior. The ceiling, proscenium and orchestra pit are designed in such a way that good sound propagation is guaranteed. The structured wooden wall linings prevent flutter echoes and help to create a good sound mix. The opera house has extensive stage machinery, with stage wagons in wings to both sides and the rear, which fulfil all possible requirements. The concert and multi-purpose halls are located beyond the side wings. This rigorous separation between the constructions enabled the necessary sound insulation to be achieved – essential if all parts of the centre are to be used at the same time. Furthermore, the restaurant complex above the auditorium is supported on special bearings to isolate it from the hall.

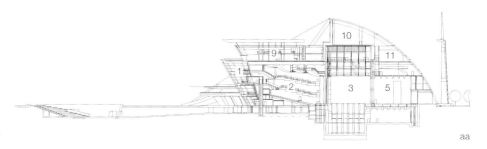

aa

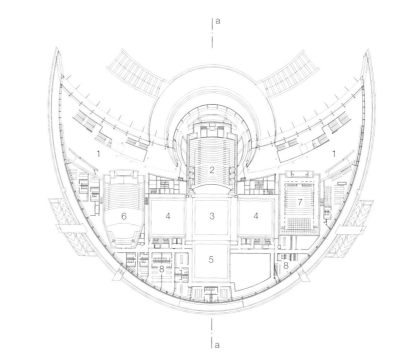

Plan · Section
scale 1:2000

1	Foyer	5	Rear wing
2	Auditorium, opera house	6	Concert hall
3	Stage	7	Multi-purpose hall
4	Side wing	8	Dressing rooms
		9	Restaurant
		10	Rehearsal stage
		11	Rehearsal room

Philharmonic Hall in Essen

Architects: Busmann + Haberer, Cologne
Acoustics
consultants: Müller-BBM, Planegg
Completed: 2004

The existing hall for the Philharmonic Orchestra in Essen was completely renovated in 2003–2004, a project that involved gutting the building and adding an extra basement storey and raising the ceiling to the hall in order to achieve the volume required for the desired reverberation. The new Alfried Krupp Hall has raking stalls seating, a balcony at the rear, choir seating and three side balconies, adding up to a total of 1860 seats. In terms of size and capacity, this hall can therefore compete with numerous well-known classical venues. The essentially rectangular form (the so-called shoebox) results in a good distribution of sound throughout the hall, with intensive early side reflections which are especially important for the three-dimensional perception of the sound. The structuring and shaping of the surfaces, and the materials used, are such that an even distribution and mixing of the sound is achieved. The polygonal panel above the stage provides the orchestra with early sound reflections and helps the musicians to hear each other properly. Furthermore, this panel also helps to direct early sound reflections into the audience. Its height is adjustable and so the transmission times of the sound reflections can be altered, which enables fine adjustment and adaptation of the acoustics to suit different orchestras and different orchestral seating arrangements.

The Alfried Krupp Hall is designed for high-quality concerts of all kinds. In addition to the wide range of classical music, for instance, electroacoustically amplified jazz concerts also take place here. The room acoustics situation in the concert hall can be adjusted by way of variable measures in the form of sound-absorbent curtains to achieve the acoustic conditions necessary for amplified live music. The hall has not only passed its acoustic

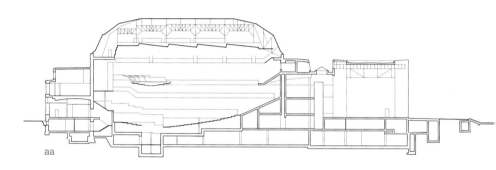

aa

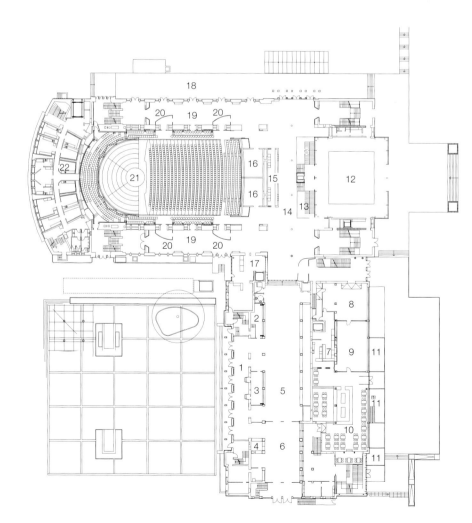

Plan of ground floor
Section   scale 1:1000
1	Entrance	5	Cloakrooms	11	Lightwell	17	Banquet servery
2	Congress office	6	Foyer restaurant	12	RWE Pavilion	18	Terrace
3	Box office	7	Kitchen	13	Void	19	Circulation zone
4	Information desk	8	Cafeteria	14	Foyer	20	Acoustic lobby
		9	Pub	15	Bar	21	Stage
		10	Rotisserie	16	Storage	22	Dressing rooms

performance test in many concerts, it has indeed been classed as outstanding by many concert-goers and active musicians.

Guaranteeing a low background noise level called for sound-insulating measures on facades, enclosing walls, roof constructions, floor constructions and entrance doors. Sources of noise outside the building such as traffic or the neighbouring underground railway were taken into account in the acoustic design. The measures used to achieve good sound insulation include supporting the entire floor beneath the stalls in the concert hall on elastic bearings and providing independent wall linings. Access to the hall is via acoustic lobbies with pairs of doors. The hall ceiling is "soundproof" and is suspended on flexible hangers to prevent the transmission of structure-borne sound.

A new, additional hall, the RWE Pavilion, has been built to complement the Philharmonic, the Alfried Krupp Hall. Acoustic aspects were taken into account during the refurbishment of and repairs to surfaces in the halls protected by preservation orders in the so-called central wing of the Philharmonic building.
All the halls are provided with curtains so that the acoustics can be varied plus mobile electroacoustic amplification systems. There are also audio and video links to enable contemporary multi-functional usage for receptions and festivals, as well as conferences and jazz concerts, while guaranteeing good acoustic conditions.

Churches

Since time immemorial, churches have been places where people go to worship God. Church architecture should therefore provide a dignified backdrop for such sacred activities.

The style of church architecture has altered considerably over the centuries and has often paved the way for new ideas in secular architecture. This has led to very diverse styles of church architecture which, all have one thing in common: their acoustic environment should endorse the sacred character of the space and convey a feeling of grandness and majesty. Only in great cathedrals can we fully appreciate the acoustics and the reverberation on such a spectacular scale. Sacred music compositions have been influenced by church acoustics for centuries, and a sustaining resonance was an indispensable prerequisite for the musical works of many ages.

Efforts to abolish the strict separation between the clergy and the congregation began during the Reformation. The sermon and the spoken word therefore increasingly became the focal point of church services.

This has resulted in the following contradictory acoustic demands for contemporary church-builders:

On the one hand, the resonance necessary for church music and suitable for the sacred nature of the interior space must be guaranteed. This is also crucial to the liturgical chants and the singing of the congregation.

On the other hand speech still needs to be readily intelligible.

It is not always possible to harmonise these conflicting requirements to everybody's satisfaction. In churches, particularly those with small congregations, the intelligibility of speech is often below standard.

Room volume and reverberation time
In churches, too, the most important room acoustics criterion is the reverberation time, which is dependent on the volume of the interior space. Expressed simply, and backed up by detailed studies and experience, we can say that for smaller churches with a volume of about 1000 m³ a reverberation time of about 2 s is satisfactory. If the reverberation time is much longer than that, the intelligibility of speech begins to suffer. Furthermore, there is also the risk that the sound of the organ will be undifferentiated and over-

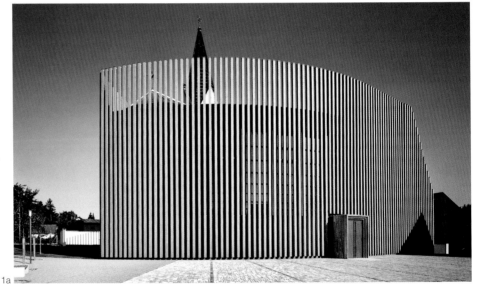
1a

b

106

2

1 a, b Extension to St. Peter's Church, Wenzenbach, 2003, Brückner+Brückner
The origins of the current church in Wenzenbach near Regensburg can be traced back to the 7th century. A large opening was created in one side wall of the church so that the interior could be extended to the side to cater for a growing congregation. The 10 m height of the extension and the 350 m² floor area ensure a good reverberation time for church services. The exposed roof beams (in a herringbone layout) and the vertical posts supporting the facade glazing scatter the sound

and take some harshness out of the reflections. The organ gallery inserted into the existing church structure opens out slightly towards the new church interior and therefore assists the radiation of the sound from the instrument.
2 Church of the Apostles, Rosenheim, 2002, Detlef Wallishauser, Thomas Krücke
Despite its circular plan form, this church has good acoustics. The recesses in the walls and the gallery at the back resolve the room geometry acoustically. In addition, the arrangement of altar, ambo and pews has been chosen well for this interior.

bearing. In a larger church with a volume of about 5000 m³ a typical reverberation time would be approx. 4 s, and with a volume of about 15 000 m³ approx. 6 s. These values apply to an empty church without upholstered pews.

Looking at the situation in more detail, we discover differences depending on the period in which the church was built and also the confession. For example, Catholic churches tend to prefer somewhat longer reverberation times to those of Protestant churches.

In order that the desired resonance is still obtained when the church is full, an adequate room volume per church-goer is an important criterion. The volume per seat in smaller churches should therefore be about 7 m³ at least and in larger churches no less than 10–15 m³.

When building a new church, a limited amount of absorbent surfaces can be incorporated in the architectural concept in order to improve the intelligibility of speech during poorly attended services. But the excessive use of absorbent surfaces should never be allowed and their use should always be coordinated with the organ design.
The acoustic properties of a church – especially with respect to speech intelligibility – with only a small congregation can be improved with the help of sound-absorbent seat cushions or lightly upholstered chairs.

Room geometry
Many different forms have been used for the interiors of churches, which also have an effect on the acoustic conditions. Generally speaking, the acoustics in circular churches are difficult to master. But there are also very good acoustic solutions such as the Church of the Apostles in

Rosenheim (Fig. 2), or the Gethsemane Church in Würzburg (completed in 2000; architect: von Branca).
In new churches attention should be given to ensuring that early reflections from the ceiling and walls reach the congregation in order to assist the natural transmission of music and speech. Essentially, the requirements and physical laws described in the previous chapter apply here as well. Owing to the comparatively long reverberation time, surfaces that cause late individual reflections are not such a disadvantage here as they are in performing arts venues with less reverberation.
In excessively reverberant church interiors, sound-screening surfaces can be helpful as well. The pulpit canopy is a well-known form of such a screen. These surfaces reduce the excitation of the room reverberation in the first place and therefore lead to a better "noise-to-sound ratio".

Sound reinforcement systems
The slow, clearly articulated manner of speaking previously common in churches is gradually disappearing for various reasons. Satisfactory intelligibility of speech is therefore becoming a key issue even in smaller churches.
Luckily, the systems available these days enable far better acoustic transmission qualities to be achieved than is the case with conventional line array loudspeakers distributed around the church, even in church interiors with intensive reverberation. In particular, narrow line array loudspeakers or well-directed individual loudspeakers guarantee a precise supply of sound to the members of the congregation while limiting the excitation of the room's reverberation. An important technical prerequisite is that the loudspeakers should not point towards the microphone positions because this increases the risk of feedback, which diminishes the sound quality and the potential amplification.

Loudspeakers should therefore not be located behind the altar or ambo, but rather to the sides or above. With a "suspended in mid-air" or central positioning, it is sometimes difficult to harmonise the acoustic and architectural aspects. But in the end, the sound reinforcement system should be regarded as an important element in the planning which – with appropriate professional assistance – should be considered at an early stage and accorded an appropriate budget. Unfortunately, in many refurbishment projects and new churches the quality of the SR systems is far below the standard achievable these days.

Induction loop systems
The needs of persons with impaired hearing should also be given due attention. By building in a so-called induction loop, e.g. in the floor, the microphone signal can be transmitted directly to the hearing aids commonly used these days.

Organ
Many different factors have to be taken into account for the design and position of the organ. It is particularly important to ensure that the sound can reach the pews and the chancel easily, which essentially means that the front of the organ (i.e. the pipes) is readily visible, not concealed in a deep alcove or aisle. The choir or soloists, should also be positioned where they can hear the sound of the organ directly. When determining the necessary dimensions of the organ, the architect should contact organ-builders or an organ specialist at an early stage when a new church is being planned.

Authorities, institutes and trade associations (selection)

Federal Environment Agency
www.umweltbundesamt.de/index-e.htm

Verein der Materialprüfungsanstalten
(association of materials-testing institutes)
www.vmpa.de

Deutsches Institut für Bautechnik
(German building technology institute)
www.dibt.de/index_eng.html

Deutsche Gesellschaft für Akustik
(German acoustics association)
www.dega-akustik.de

European Acoustics Association
www.european-acoustics.org

Institute of Noise Control Engineering
of the USA (INCE/USA)
www.inceusa.org/index.asp

Bibliography (selection)

Beranek, L.: Opera and Concert Halls –
How they sound, 2nd ed., Springer Verlag,
New York, 2004

Eggenschwiler, K.: Aktuelle Aspekte der
Kirchenakustik, Schweizer Ingenieur und
Architekt, No. 25, 25 June 1999, pp. 8–12

Fasold, W., Veres, E.: Schallschutz und Rau-
makustik in der Praxis, 2nd ed.,
Verlag für Bauwesen, 2003

Gösele, K., Schüle, W., Künzel, H.: Schall,
Wärme, Feuchte, 10th ed. Bauverlag, 1997

Huber, L., Kahlert, J., Klatte, M. (ed.): Die
akustisch gestaltete Schule: Auf der Suche
nach dem guten Ton. Edition Zuhören vol. 3,
Vandenhoeck & Ruprecht, 2002

Izenour, G. C.: Theatre Design, 2nd ed.,
Yale University Press, 1997

Kuttruff, H.: Room Acoustics, 4th ed.,
Spon Press, London, 2000

Meyer, J.: Akustik und musikalische Auffüh-
rungspraxis, 4th ed., Edition Bochinsky, 2004

Meyer, J.: Kirchenakustik, Edition Bochinsky
2003

Mommertz, E., Engel, G., Drescher, K.: Besser
leise lernen, TrockenbauAkustik, 11/2002

Müller, G., Möser, M.: Taschenbuch der
Technischen Akustik, 3rd ed., Springer
Verlag, Berlin, 2004

Ruhe, C.: Schallschutz von Haustrennwänden
– Die Fuge macht's. Beratende Ingenieure,
2003

Schönwälder, H.-G., Berndt, J., Ströver, F.,
Tiesler, G.: Lärm in Bildungsstätten – Ursachen
und Wirkung, Federal Institute for Occupa-
tional Safety & Health, Fb 1030, NW-Verlag,
2004

Schricker, R.: Kreative Raum-Akustik für
Architekten und Designer, DVA Stuttgart,
2001

Standards and directives (selection)

DIN 4109: Sound insulation in buildings;
requirements and testing, Nov 1989
(with supplements and amendments)

DIN 18005-1: Noise abatement in town
planning – Part 1: Fundamentals and
directions for planning, Jul 2002

DIN 18005 supp. 1: calculation methods;
acoustic orientation values in town planning,
May 1987

DIN 18041: Acoustic quality in small to
medium-sized rooms, May 2004

DIN 45691: Noise allotment, Dec 2006

DIN EN 12354 parts 1 – 6: Building acoustics
– Estimation of acoustic performance of
buildings from the performance of elements,
2000 – 2007

DIN EN 29053: Acoustics; materials for
acoustical applications; determination of airflow resistance, May 1993

DIN EN ISO 140 parts 1, 3 – 8, 11, 14, 16, 18:
Acoustics – Measurement of sound insulation
in buildings and of building elements,
2000 – 2007

DIN EN ISO 354: Acoustics – Measurement
of sound absorption in a reverberation room,
2003

DIN EN ISO 3382: Acoustics – Measurement
of room acoustic parameters, Mar 2000

DIN EN ISO 10534 parts 1 & 2: Acoustics –
Determination of sound absorption coefficient
and impedance in impedances tubes,
Oct 2001

DIN EN ISO 10848 parts 1 – 3: Acoustics –
Laboratory measurement of the flanking
transmission of airborne and impact sound
between adjoining rooms, 2006

DIN EN ISO 11654: Acoustics – Sound absorbers for use in buildings – Rating of
sound absorption, Jul 1997

VDI 2081 sheet 1: Technical rule – Noise
generation and noise reduction in air-condi-
tioning systems, Jul 2001

VDI 2566 sheet 1/2: Technical rule – Acoustical
design for lifts with/without a machine room,
2001/2004

VDI 2569: Technical rule – Sound protection
and acoustical design in offices, Jan 1990

VDI 2719: Technical rule – Sound isolation
of windows and their auxiliary equipment,
Aug 1987

VDI 3728: Technical rule – Airborne sound
isolation of doors and movable walls,
Nov 1987

VDI 4100: Technical rule – Noise control in
dwellings – Criteria for planning and assessment, Aug 2007

Directive for Noise Abatement on Roads
(RLS-90), Federal Ministry of Transport,
Bonn, 22 May 1990, reprinted with correc-
tions Feb 1992

Directive for the calculation of sound emissions from rail traffic (Schall 03), Federal
Railways Central Office, Munich, 1990

UK Building Regulations, Approved Document E "Resistance to the passage of
sound", 2003 ed.

Manufacturers (selection)

Sound-absorbent surfaces

Akustik Plus GmbH
www.akustik-plus.com

Akustik & Raum AG
www.akustik-raum.com

Armstrong Building Products GmbH
www.armstrong.com

Armstrong Metalldecken AG
www.gema.biz

Caparol
www.caparol.de

Clauss markisen Projekt GmbH
www.clauss-markisen.de

Création Baumann GmbH
www.creationbaumann.com

CS-Interglas AG
www.cs-interglas.com

Danogips GmbH & Co KG
www.danogips.de

Knauf Insulation GmbH & Co. KG
www.heraklith.com

Deutsche Rockwool Mineralwolle
GmbH & Co. OHG
www.rockfon.de

Diel Absorber
www.diel-absorber.de

Eurofoam
www.euro-foam.com

Franz Habisreutinger GmbH & Co. KG
www.habisreutinger.de

Fural Systeme in Metall GmbH
www.fural.at

Gerriets GmbH
www.gerriets.de

Girnghuber GmbH
www.gima-ziegel.de

Gruber Technik
www.spaceartfactory.com

Henkel Bautechnik GmbH
www.phonestop.de

Integrale Climasysteme GmbH
www.kuehldecke.de

KAEFER Construction GmbH
www.microsorber.de

Knauf AMF GmbH & Co. KG
www.amfgrafenau.de

Knauf Gips KG
www.knauf.de

Koch Membranen GmbH
www.koch-membranen.de

Lahnau Akustik GmbH
www.lahnau-akustik.de

Lignokustik AG
www.lignokustik.ch

Lignotrend Produktions GmbH
www.lignotrend.com

Lindner AG
www.lindner-holding.de

Modul Betonstein GmbH + Co. KG
www.modul-betonstein.de

Muhlack Kiel GmbH
www.muhlack.de

Normalu-Barrisol S.A.
www.barrisol.de

Odenwald Faserplattenwerk GmbH
www.owa.de

Pinta acoustic GmbH
www.pinta-acoustic.de

rohi stoffe GmbH
www.rohi.de

Saint Gobain Rigips GmbH
www.rigips.de

Saint-Gobain Decoustics AG
www.decoustics.ch

Saint-Gobain Ecophon GmbH
www.ecophon-international.com

Schmitz-Werke GmbH + Co. KG
www.drapilux.de

Sto Verotec GmbH
www.stoverotec.de

Texaa designing silence
www.texaa.com

Verotex AG
www.verotex.de

Zent Frenger
www.zent-frenger.de

Sound insulation for floating floor finishes

BSW GmbH Berleburger Schaumstoffwerk
www.berleburger.de

Calenberg Ingenieure GmbH
www.calenberg-ingenieure.de

Dt. Rockwool Mineralwoll GmbH & Co. OHG
www.rockwool.de

Getzner Werkstoffe GmbH
www.getzner.com

Gutex Holzfaserplattenwerk GmbH + Co. KG
www.gutex.de

Homatherm GmbH
www.homatherm.com/uk

Knauf Insulation GmbH & Co. KG
www.heraklith.com

Pavatex GmbH
www.pavatex.de

Philippine GmbH & Co. Dämmstoffsysteme KG
www.philippine-eps.de

Rockwool Ltd.
www.rockwool.co.uk

Saint-Gobain Isover G+H AG
www.isover.de

Schöck Bauteile GmbH
www.schoeck.de

Schwenk Dämmtechnik GmbH & Co. KG
www.schwenk-daemmtechnik.de

Thermal Ceramics Deutschland GmbH, Co. KG
www.tc-sitek.com

Unidek Deutschland GmbH
www.unidek.de

URSA Deutschland GmbH
www.ursa.de

Noise control doors, movable room dividers

Becker GmbH & Co. KG
www.becker-tw.de

Buchele GmbH
www.buchele.de

Dorma Hüppe Raumtrennsysteme GmbH + Co. KG
www.dorma-hueppe.com

Franz Nüsing GmbH & Co. KG
www.nuesing.com

Hodapp GmbH & Co. KG
www.hodapp.co.uk

IAS Industrie Akustik Siegburg GmbH
www.ias-tueren.de

JELD-WEN Deutschland GmbH & Co. KG
www.wirus.de, www.moralt.de

Schörghuber Spezialtüren KG
www.schoerghuber.de

Svedex B.V.
www.svedex.com

Noise control glazing and windows, facades

Hydro Building Systems GmbH
www.wicona.ch

Pilkington Holding GmbH
www.pilkington.com

Saint-Gobain Glass Deutschland GmbH
www.saint-gobain-glass.com

Schüco International KG
www.schueco.com

Low-noise ventilation fans and sound-insulated make-up flow inlets

emco Bau -und Klimatechnik GmbH & Co. KG
www.emco-klima.de

LTG AG
www.ltg-ag.de

Trox GmbH
www.trox.de

Westaflexwerk GmbH
www.westaflex.com

Noise barriers

BECK Sound Barrier Systems, Inc.

www.beck-soundbarriers.com

Betonwerk Rieder GmbH
www.rieder.at

Finnforest UK
www.finnforest.co.uk
Franken Schotter GmbH & Co. KG
www.franken-schotter.de

K. Schütte GmbH
www.schuette-aluminium.de

Index

absorbent backing 22, 23, 59, 61, 71, 77, 83, 88, 98, 100
absorbent lining 19–23, 47, 62, 64, 69, 71, 83
absorbent surface 14–17, 59– 63, 66, 69, 71, 72, 76–80, 83, 95, 107
absorber 20, 22, 23, 79, 100
absorption 12, 15–23, 26, 27, 45, 59, 62, 64, 69, 71, 80, 82, 83, 91, 92, 100
absorption surface area 15, 16, 17, 23, 26, 27, 59, 62, 69, 71, 82
acoustic attenuation 15, 59, 61, 63, 70, 83, 98
air traffic 39, 41
airborne sound 25–27, 29, 30, 33, 35, 51–53, 55, 61, 62, 72, 73, 101
airborne sound excitation 25
assessment level 40–43, 46
assessment time 41, 42
auditorium 67, 75, 80, 92, 94–97, 100, 103
auralisation 18
A-/C-/D-weighting 10
background noise level 59–62, 78, 105
building services 10, 25, 36, 37, 49, 50, 52, 55, 72, 77, 82, 101, 102
ceiling reflection 94, 95, 97
centre frequencies 9
clarity C_{80} 14
coincidence frequency 30, 31
computer simulation 17–19
damping 21, 23, 26, 27, 30–32, 34, 45, 59, 71, 78, 82, 92, 96, 97
decibel 9, 11, 26
definition D 14
diffraction angle 45
diffuse sound 13, 15, 16, 17, 27, 59
direct sound 12–14, 16, 17, 45, 78, 90, 91, 95, 97
directional characteristic 11, 78, 93
dry construction 28, 30, 31, 37, 51, 52, 61, 72, 74, 84, 89
dynamic stiffness 33, 34, 74, 86
early reflection 12, 13, 16, 70, 78, 90, 94, 98, 107
electroacoustic amplification 78, 79, 90, 93, 96, 97, 100, 105
electronic room acoustics 98, 99, 100, 101
emissions point 42
enhanced sound insulation 49–53, 60, 61
Environmental Noise Directive 39, 42
external noise 35, 36, 39, 42, 43, 46, 47, 54, 65, 68, 72, 74–77, 84
feedback 7, 78, 82, 100, 107
flanking component 25–28 , 30, 32, 33, 52, 55, 61, 72, 73
flanking path 27, 55
flanking sound level difference 26, 31
flanking transmission 26–29, 31, 35, 46, 47, 52, 61, 65, 72, 88
flutter echo 13, 16, 69, 71, 77, 80, 83, 86, 98, 103

frequency 8–11, 14, 15, 17–22, 25, 26, 27, 30–34, 36, 37, 46, 59, 69, 76, 78, 79, 84, 92, 93, 101, 102
frequency band 9, 10, 15, 92
health 7, 9, 10, 39, 41, 49, 54, 82, 83
human ear 8–11, 18, 26, 78
interaural cross-correlation coefficient 14
immissions limit value 40
immissions point 41–44
impact sound 25–28, 33–35, 50–56, 59, 60, 62, 64, 65, 72, 74, 77, 79, 83, 86, 89, 101
impact sound insulation 26, 27, 33–35, 52–56, 62, 65, 72, 74, 77, 86, 89
individual reflection 12, 13, 96, 107
interaural time difference 11–13
loudness level 9, 10, 16, 17, 27, 63, 82, 83, 96
loudspeaker 11, 18, 32, 44, 59, 60, 93, 96, 97, 100, 107
measurements on models 18, 19
mineral-fibre insulating material 20, 22, 31, 34
minimum sound insulation 25, 35, 49, 51, 52, 54, 55, 62, 79
noise abatement 39, 40–42, 44
noise barrier 45, 54
noise exposure 39, 40, 42, 43, 46, 56, 74
noise pollution 39, 40
noise quota 44
octave 8–10, 14, 15, 19, 20, 26, 30, 59, 68, 69, 78
one-third octave band 9, 19, 26
orchestra pit 94–98, 103
orchestra shell 94, 96, 98
party floor 52
party wall 27, 30, 31, 50–54
passive noise control 40, 41, 47, 49, 50, 84, 101
proscenium 95–98, 103
rail traffic 39–42, 101
receiving position 17
receiving room 25–27, 29, 60
reflective surface 20, 22, 42, 78, 83, 95, 96, 97, 98, 100
refurbishment 18, 19, 34, 41, 47, 52, 68, 69, 70, 71, 72, 74, 97, 105, 107
resonance 18, 21, 29, 30, 34, 36, 51, 52, 71, 88, 92, 100, 106, 107
resonant frequency 18, 21, 30, 31, 33, 34, 36, 84
resonant system 30, 33
reverberation 12–15, 17–20, 23, 27, 54, 55, 58, 59, 62–64, 68–71, 77, 78, 83, 88, 90, 91, 93, 94, 95, 96, 97, 98, 100, 104, 106, 107
reverberation time 13–15, 17–19, 27, 54, 55, 59, 62–64, 68, 69, 70, 71, 78, 83, 90, 91, 93, 94, 95, 96, 98, 100, 106, 107
road traffic 10, 36, 39–42

room acoustics measurements 17, 18
room form 90, 94, 97
room geometry 13, 17, 46, 107
room impulse response 12, 17, 18
room volume 14, 15, 17, 68, 73, 78, 83, 88, 90, 93, 94, 96, 98, 100, 106, 107
Sabine reverberation equation 15, 19
scatter 17, 18, 19, 28, 83, 107
screening 45, 49, 59–61, 95, 107
secondary airborne sound 25, 101
separating wall 27, 30, 31, 34, 55, 73, 81
sleep 10, 39, 47, 54
sound absorption coefficient 15, 17, 19, 20, 22 , 23, 62
sound bridge 32, 33, 35
sound energy 9–12, 14, 15, 19, 20, 90, 95
sound insulation 7, 25–37, 40, 45–47, 49–56, 59–62, 64, 65, 72–75, 77, 79, 82–86, 88, 89, 101, 102, 103, 105
sound insulation category 49–51, 53, 83, 84
sound insulation class 35, 36, 47, 74
sound mix 17, 20, 83, 90, 93, 94, 103
sound power level 11, 16, 27, 37, 79
sound pressure 8–11, 16, 17, 18, 26, 27, 32, 36, 37, 64
sound pressure level 9–11, 16, 17, 18, 26, 27, 32, 36, 37, 64
sound propagation 7, 11–13, 15, 18, 25–28, 34, 37, 42, 45, 59, 64, 71, 101, 103
sound reduction index 26–32, 34–36, 47, 50, 52, 53, 54, 55, 60–62, 64, 65, 73, 79, 83
sound reinforcement system 11, 78, 79, 93, 101, 107
sound signal 8–11, 13, 14, 100
sound source 11–14, 32, 41, 42, 45, 95
sound-absorbent lining 19, 22, 23, 47, 69
source room 25, 27
spectrum adaptation term 36, 47
speech 7–10, 12–15, 18, 49, 54, 59, 60, 62, 68, 69, 76, 78–80, 90, 94, 96–98, 106, 107
speech transmission index 14, 69
stage acoustics 93, 98
stairs 26, 35, 51, 53, 54, 60, 62, 64, 67
strategic noise map 39
strength G 14
structure-borne sound 9, 25, 26, 32–35, 37, 52, 53, 64, 101, 102, 105
suspended ceiling 19, 26, 30, 59, 61, 62, 69, 70, 71, 74, 86, 88, 89, 92
suspended floor 25, 26, 28, 33, 34, 35, 52, 59, 65, 74, 84
traffic noise 10, 36, 39–41, 42, 47
urban planning 10, 39–47, 65, 102
variable surface 98, 100
vibration 8, 12, 25, 30, 34, 101
walking 25, 28, 34, 49
weighted normalised impact sound level 26, 27, 28, 35, 60, 62, 64
weighted standardised sound level difference 27, 55

Picture credits

The authors and publishers would like to express their sincere gratitude to all those who have assisted in the production of this book, be it through providing photos or artwork or granting permission to reproduce their documents or providing other information. All the drawings in this book were specially commissioned. Photographs not specifically credited were taken by the architects or are works photographs or were supplied from the archives of the magazine DETAIL. Despite intensive endeavours we were unable to establish copyright ownership in just a few cases; however, copyright is assured. Please notify us accordingly in such instances.

page 6:
Mark Fiennes, arcaid/archenova, London

page 18, 19, 20, 23, 26, 29, 32, 33, 44, 45, 47, 59 left, 71 right, 73, 82, 87, 92, 94 left and centre, 99 top:
Müller-BBM, Planegg

page 24, 48, 66, 67, 80, 81 left:
Frank Kaltenbach, Munich

page 35:
Finnforest Merk GmbH, Aichach

page 38:
Franken-Schotter GmbH & Co KG,
Treuchtlingen-Dietfurt

page 46:
Melanie Schmid, Munich

page 52:
Oliver Schuster, Stuttgart

page 56, 57, 79 top left:
Michael Heinrich, Munich

page 58:
Jens Heilmann/KMS TEAM GmbH,
Munich

page 59 right:
Gesellschaft für Akustik & Gestaltung mbH,
Bietigheim-Bissingen

page 60 left:
renz solutions GmbH, Aidlingen

page 60 right:
Richie Müller, Munich / feco Innenausbausysteme GmbH, Karlsruhe

page 63:
Robert Deopito, Vienna

page 69 left:
Knauf AMF GmbH & Co. KG, Grafenau

page 69 right:
Heraklith GmbH

page 70 top, 71 left:
Knauf Gips KG, Iphofen

page 75, 76, 77:
Gerhard Hagen, Bamberg

page 79 top right:
Nigel Young / Foster & Partners

page 79 bottom, 108:
Christian Gahl, Berlin

page 81 right:
Thomas Mayer Archive, Neuss

page 85, 86:
Richie Müller, Munich

page 88, 89:
Rolf Sturm, Landshut

page 90:
Gesellschaft der Musikfreunde Collection,
Vienna

page 91:
VG Bild-Kunst, Bonn

page 93:
Ulrich Schwarz, Berlin

page 94 right:
Karl Forster / Salzburger Festspiele

page 97:
Andreas Pohlmann, Munich

page 98 right:
Siegfried Wameser, Munich

page 99 bottom:
Franziska von Gagern, Munich

page 101:
KKL Luzern Management AG

page 106:
Peter Manev, Selb

Full-page plates

page 6:
Roman amphitheatre in Orange, France
The impressive thing about an amphitheatre is that you can hear every word perfectly even when sitting in the back row. However, this does call for a low level of background noise, which means that sometimes nearby streets have to be cordoned off during events.

page 24:
Swiss Re offices in Munich, 2002
Architects: BRT Architekten, Hamburg
Movable room dividers enable full use to be made of the space available or certain areas to be visually and acoustically screened off to the suit the exact requirements of users.

page 38:
Noise barrier in Ingolstadt, 2005
Stone-filled wire cages (gabions) are being increasingly favoured as noise barriers. A very specific layered construction ensures the necessary sound absorption and attenuation.

page 48:
Perimeter block housing in Rotterdam, 2002
Architects: KCAP, Rotterdam
The glazed loggias give the facade a very distinctive appearance, create additional space for occupants and also serve as a buffer against exterior noise.

page 108:
In acoustic testing laboratories (also called anechoic chambers), sound-absorbent wedges 1 m deep and more suppress all reflections. A feeling of pressure in the ears is often experienced when entering such a room.